GEOLOGY OF THE SAN DIEGO METROPOLITAN AREA, CALIFORNIA

GEOLOGY OF THE SAN DIEGO METROPOLITAN AREA, CALIFORNIA

BY

Michael P. Kennedy

KALPAZ

ISBN: 978-93-5128-619-6(HB)

First Published in 2001
Indian Reprint in 2017

Published by

Kalpaz Publications
C-30, Satyawati Nagar,
Delhi – 110052
E-mail: kalpaz@hotmail.com
Website: kalpazpublications.com
Ph.: +91-9212142040

Cataloging in Publication Data—DK
Courtesy: D.K. Agencies (P) Ltd. <docinfo@dkagencies.com>

Kennedy, Michael P., author.
Geology of the San Diego Metropolitan Area, California / by Michael P. Kennedy.
pages cm
"First published in 2001"—Title page verso.
Includes bibliographical references.
Contents: section A. Western San Diego Metropolitan Area — section B. Eastern San Diego Metropolitan Area.
ISBN 9789351286196 (HB)

1. Geology—California—San Diego Metropolitan Area. I. Title.

QE90.S116K46 2017 DDC 557.9498 23

CONTENTS

ILLUSTRATIONS

SECTION A – WESTERN SAN DIEGO METROPOLITAN AREA

SECTION B – EASTERN SAN DIEGO METROPOLITAN AREA

SECTION A

Del Mar, La Jolla, and Point Loma quadrangles

ABSTRACT

The Del Mar, La Jolla, and Point Loma quadrangles are underlain by sedimentary rocks of Late Cretaceous, Eocene, Pliocene, Pleistocene, and Holocene age that rest with angular unconformity on a Mesozoic metamorphic and plutonic rock basement complex. The Tertiary and Quaternary sedimentary succession was deposited unconformably on the Upper Cretaceous strata in a northwest-trending basin herein referred to as the San Diego embayment. The most abundant rocks of the San Diego embayment are gently folded and faulted Eocene marine, lagoonal and nonmarine rocks that form a northwest-trending, eastward-thinning section. These strata were laid down upon the older rocks during a period of regional tectonic downwarping.

The pre-Eocene rocks of the area from oldest to youngest belong to the Upper Jurassic Santiago Peak Volcanics, mid-Cretaceous southern California batholith, and the Upper Cretaceous Rosario Group. The Santiago Peak Volcanics are mildly metamorphosed and occur in the subsurface throughout most of the southern California continental margin but are exposed in only a few places within this area. The rocks consist of interlayered meta-andesite, meta-quartz latite, meta-shale, tuff, slate, and quartzite. A meager marine molluscan fauna from the meta-shale indicates a Late Jurassic (Portlandian) age. The Santiago Peak Volcanics are intruded by rocks of the southern California batholith. The batholithic rocks that crop out in the mapped area are mostly gabbros. These rocks are locally deeply weathered and difficult to distinguish from overlying nonmarine Eocene strata that have been largely derived from detritus of plutonic origin. Deformation, uplift, and unroofing of the batholith occurred prior to the deposition of the Upper Cretaceous clastic marine and nonmarine strata of the Rosario Group. The basal formation of the Rosario Group is the Lusardi Formation, a nonmarine boulder fanglomerate that was deposited along the western margin of the tectonic highlands upon the weathered surface of the plutonic and metamorphic rock. Clasts in the Lusardi Formation are composed of locally derived basement rocks. Good exposures of the Lusardi Formation occur approximately 16 kilometers (km) north of Del Mar and 6 km east of Carlsbad where the conglomerate is overlain by the middle part of the Rosario Group, marine sandstone and siltstone of the Point Loma Formation. The Point Loma Formation is Campanian and Maestrichtian in age and underlies most of the Point Loma Peninsula and the hills southeast of La Jolla. It is conformably overlain by the uppermost part of the Rosario Group, marine sandstone and conglomerate of the Maestrichtian Cabrillo Formation.

Nine partially intertonguing middle and upper Eocene formations composed of siltstone, sandstone, and conglomerate were deposited during two major transgressive-regressive cycles upon an erosional surface of mild relief following uplift and erosion of the Upper Cretaceous strata. The succession is in excess of 700 meters (m) thick and grades from nonmarine fan and dune deposits on the east through lagoonal and nearshore beach and beach-bar deposits to marine continental shelf deposits on the west near the present-day coastline. The age and environmental interpretation of the rocks is based on the mapped distribution of the lithofacies and by presence of fossil calcareous nannoplankton and foraminifers in the continental shelf succession, mollusks in the nearshore rocks, and vertebrate land animals in the nonmarine sequence. Flame structures and current ripple marks in the continental shelf deposits, cross bedding in the nearshore deposits, and cobble imbrications and paleo-stream gradients in the deltaic, lagoonal, and fluviatile deposits combined with their petrologic content indicate that the sediments were derived from local source areas to the east.

The nonmarine facies of the Eocene formations are typically well indurated and cemented. The lagoonal facies are soft, friable, and poorly cemented. The nearshore facies are well indurated, well sorted, and locally concretionary. The marine deposits are typically fine grained, well indurated, and well cemented.

Rocks of the Pliocene San Diego Formation, where preserved, rest unconformably upon the Eocene strata. The San Diego Formation is in turn overlain by the Lindavista Formation, a combination of nearshore marine, beach, and nonmarine strata composed mostly of sandstone and conglomerate. The Lindavista Formation was deposited on a broad wave cut terrace that extends across the entire width of the area. The late Pleistocene Bay Point Formation and Holocene surficial deposits complete the stratigraphic record.

Tectonic deformation within the area can be divided into two episodes: (1) pre-mid-Cretaceous, and (2) late Tertiary and Quaternary. The Santiago Peak Volcanics were chaotically deformed and partly overturned during the first episode. The less deformed rocks that have been faulted and gently folded by the later episode include those of Upper Cretaceous and later age. Sediments approximately 100,000 years in age have been vertically offset in excess of 20 m by youthful faults that transect the area. The most prominent of these include the Rose Canyon, Mount Soledad, Old Town, and Point Loma faults. Speculation has been made in recent literature that the Rose Canyon fault is related to the active Newport-Inglewood structural zone on the north and the San Miguel fault in northern Baja California on the south. Forty-four earthquakes of Richter magnitudes between 2.5 and 3.7 (M 2.5 and M 3.7) and having epicentral localities within the greater San Diego area have been recorded by the California Institute of Technology Seismological Laboratory since 1950. It has been shown that the area has had a strain release of between 1 and 16 equivalent magnitude 3 earthquakes/400 km^2 for the 29-year period between 1934 and 1963.

Seismically triggered landslides have occurred in the sea cliffs at Point Loma, La Jolla, and Torrey Pines. Most of the mapped landslides, however, are gravity slides attributable to soft incompetent material, ground water penetration, and oversteepened slopes.

Sand and gravel deposits useable for concrete, bituminous, and ceramic aggregate underlie a large part of the area. Clay deposits useable for ceramics, fire clay, and expansible clay are also abundant but have not been exploited. The clay deposits are widespread and closely associated with expansive soils and surficial landsliding.

Photo 1. The San Diego coastal area and adjacent Peninsular Range Province showing boundaries of the Del Mar, La Jolla, and Point Loma quadrangles.

GEOLOGY OF THE WESTERN SAN DIEGO METROPOLITAN AREA, CALIFORNIA

Del Mar, La Jolla, and Point Loma quadrangles

by Michael P. Kennedy[1]

INTRODUCTION

In 1965 the California Division of Mines and Geology in cooperation with the City of San Diego began a comprehensive geologic investigation aimed at a better understanding of the geologic hazards that exist within the greater San Diego metropolitan area (Kennedy, 1967, 1969). This report is one product of that investigation and is complemented by a similar report on the La Mesa, Poway, and SW1/4 Escondido quadrangles (Kennedy and Peterson, 1975). Together the Del Mar, La Jolla, and Point Loma quadrangles are approximately 350 square kilometers (km^2) in extent and constitute the western part of the greater San Diego metropolitan area (figure 1 and photo 1).

The western San Diego metropolitan area is underlain by valuable sand, gravel, and clay resources deemed feasibly extractable in today's market for use in the north county, Del Mar, La Jolla, Miramar, Lindavista and Point Loma areas. The area is underlain primarily by sedimentary rock; however, occasional outcrops of plutonic and metamorphic rocks do occur. Very small surficial landslides (mostly unmapped due to scale of map) associated with expansible clay deposits in the northern and eastern parts of the area are abundant. These landslides are closely associated with the outcrops of Friars and Delmar Formations. The rock units mapped and discussed herein are shown in diagrammatic relationship in figure 2.

Previous investigations that have been especially useful in this study include a ground water investigation by A.J. Ellis (1919), a stratigraphic and paleontologic thesis of the La Jolla quadrangle by M.A. Hanna (1926), studies of the Pliocene deposits of San Diego by L.G. Hertlein and U.S. Grant IV (1939, 1944), a monograph on the mineral resources of San Diego County by F.H.

[1]Geologist, California Division of Mines and Geology

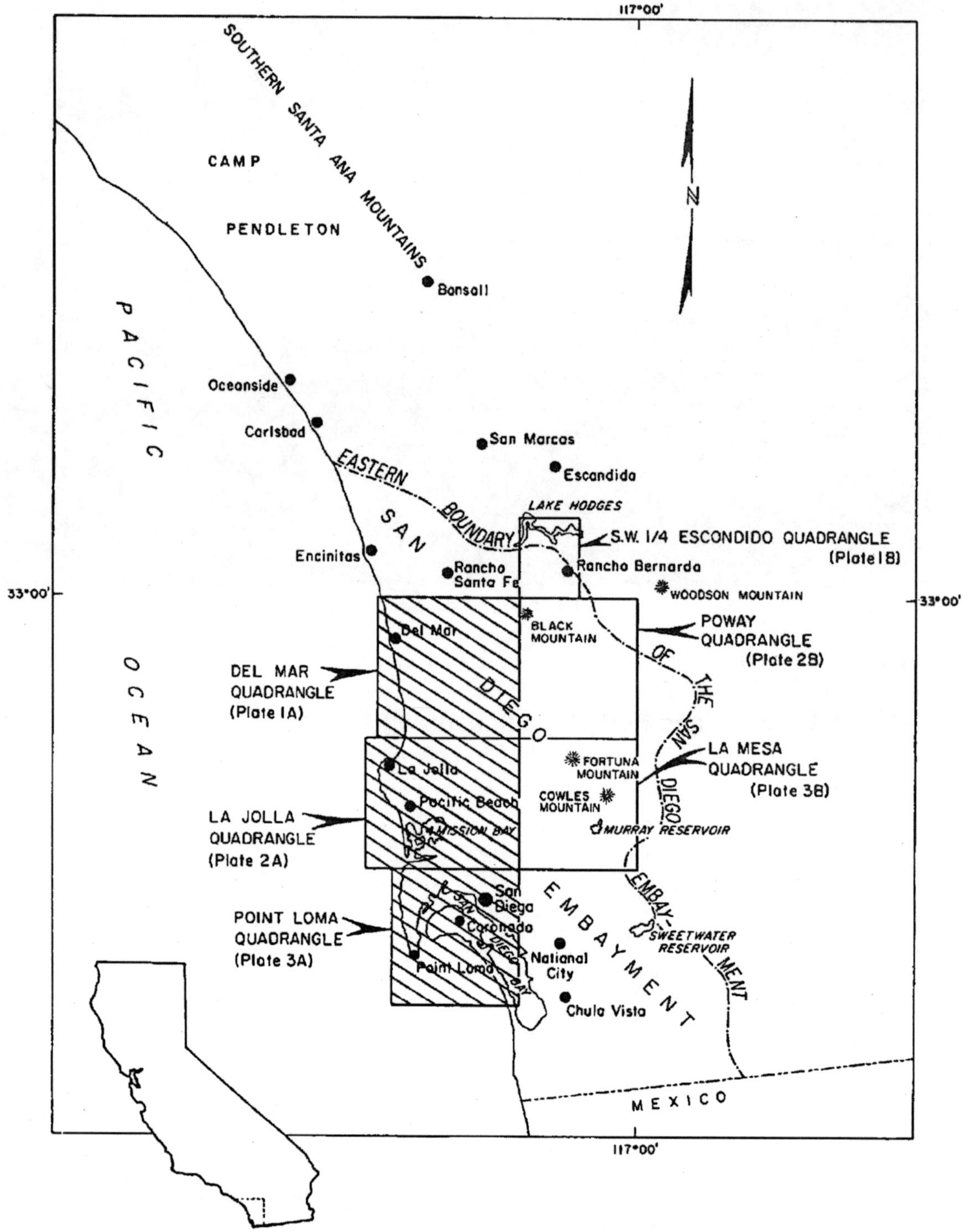

Figure 1. Index map showing the location of the Del Mar, La Jolla, and Point Loma quadrangles.

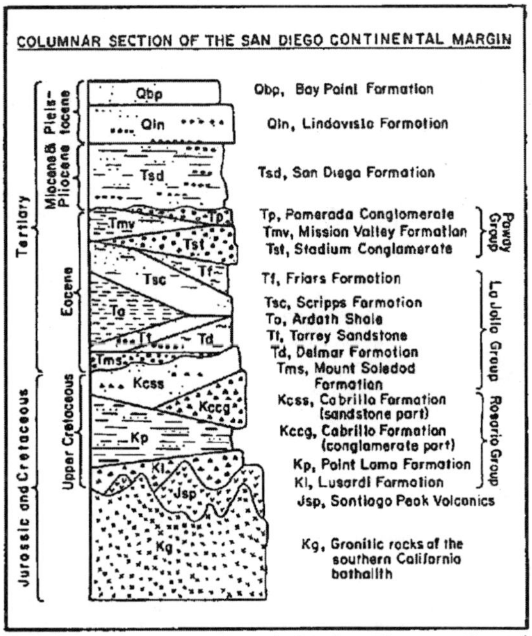

Figure 2. Columnar section of the San Diego continental margin.

Weber (1963), and the San Diego-El Centro geologic map sheet by R.G. Strand (1962).

The author would like to extend special thanks to D.M. Morton and G.W. Moore of the United States Geological Survey for encouragement, help in the field, and many valuable discussions pertinent to this study. Acknowledgment is due also to M.O. Woodburne and M.A. Murphy of the University of California Riverside and Professor A.O. Woodford of Pomona College for long interest in this study and enthusiastic help in the field and laboratory; to D. Bukry, D.J. Golz, C.R. Givens, J.P. Kern, E.D. Milow, W.J. Zinsmeister, and the late E.C. Allison for assistance in the paleontologic aspects; A.K. Baird for help in petrographic aspects; P.K. Morton, C.H. Gray, Jr., G.B. Cleveland, B.W. Troxel, F.H. Weber, Jr., Y.H. Smitter, R.G. Strand, G.L. Peterson, and J.I. Ziony for many interesting discussions in the field and for reviewing the maps and manuscript.

GEOLOGIC SETTING

Pre-Eocene rocks in the southern Peninsular Ranges of California are subdivided into four major units. From oldest to youngest they include the Bedford Canyon Formation, Santiago Peak Volcanics, southern California batholith, and the Rosario Group. Together these units form the basement complex upon which the younger sedimentary succession rests. See figure 2 and cross sections A-A' (plate 1A), B-B', C-C' (plate 2A), and D-D' (plate 3A).

The Santiago Peak Volcanics (Black Mountain Volcanics of Hanna, 1926) rest with angular unconformity on the Bedford Canyon Formation where the latter has been preserved. The Bedford Canyon Formation is not known to exist at the surface in this area. In the Santa Ana Mountains to the north, the Santiago Peak Volcanics have an exposed length of 130 km (Larsen, 1948), and to the south they extend from the international boundary to near the center of Baja California (Allison, 1964). They occur in the subsurface throughout most of the southern California continental margin (Hertlein and Grant, 1944; Gray *et al.*, 1971).

The Santiago Peak Volcanics have undergone mild metamorphism and have been intruded by rocks of the southern California batholith. The plutonic rocks of the batholith that crop out in the mapped area are gabbros, which have a steeply inclined contact with the older metamorphic rock.

The southern California batholith forms the backbone of the Peninsular Ranges of southern California and Baja California and is nearly 1,500 km in length extending from the Transverse Ranges on the north to the southern part of the Baja California peninsula on the south. The batholithic rocks within the study area were named and described by Larsen (1948).

Deposition of Upper Cretaceous clastic marine and nonmarine strata followed the emplacement, uplift, unroofing, and deformation of the southern California batholith (figure 3). The basal formation of the clastic succession is the Lusardi Formation, a nonmarine boulder fanglomerate that forms the base of the Rosario Group. The Lusardi Formation was laid down along the western margin of the tectonic highlands and upon a deeply weathered surface of the plutonic and metamorphic rock (Peterson and Nordstrom, 1970). The clasts of the Lusardi Formation are composed essentially of these two rock types, suggesting a local source area (Nordstrom, 1970). The Point Loma Formation is the intermediate formation of the Rosario Group. It underlies most of the Point Loma Peninsula and the hills southeast of La Jolla and is conformably overlain by marine sandstone and conglomerate of the Cabrillo Formation. The Cabrillo Formation is the uppermost formation of the Rosario Group and is also exposed at Point Loma and La Jolla.

The pre-Eocene basement terrain is locally decomposed to depths of 50 meters. In most areas where Eocene rock rests directly on the basement rock, the early Tertiary surface (sub-La Jolla unconformity, figure 3) is marked by residual clay deposits of montmorillonite that grade downward to fresh basement rock and upward into the Eocene sedimentary rock. The decomposed granitic rock and clay were primary sources of sediment for the Eocene depositional basin and give rise to the granitic appearance of the arkosic sandstone of these sedimentary facies.

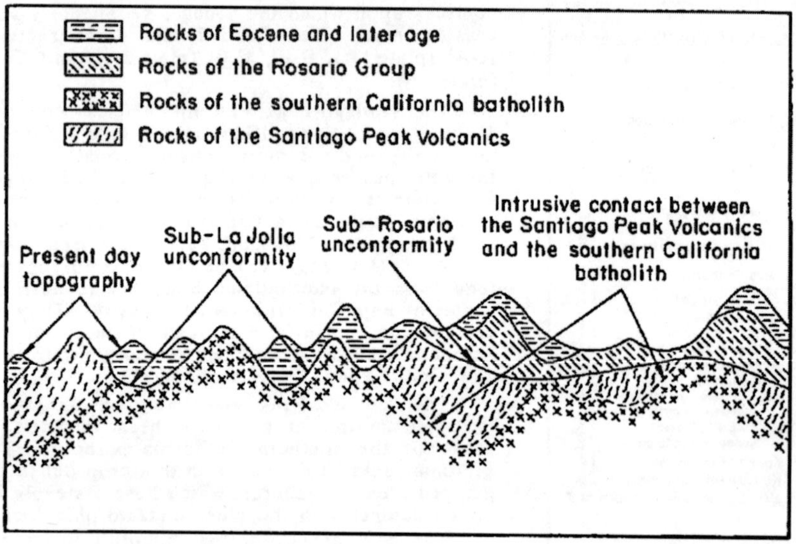

Figure 3. Diagrammatic sketch of the basement complex and superjacent strata.

PRE-EOCENE DEPOSITS

Santiago Peak Volcanics

The Santiago Peak Volcanics comprise an elongate belt of mildly metamorphosed volcanic, volcaniclastic, and sedimentary rocks that crop out from the southern edge of the Los Angeles basin southward into Mexico. They were originally named "Black Mountain Volcanics" (Hanna, 1926, p. 199-204) for exposures in the northeast part of the area. Larsen (1948) substituted the name—Santiago Peak Volcanics—as the name "Black Mountain" was pre-empted.

The volcanic rocks range in composition from basalt to rhyolite but are predominantly dacite and andesite. The succession is typified by a wide variety of breccia, agglomerate, volcanic conglomerate, and fine-grained tuff and tuff breccia. Highly silicified rock—probably tuff and a variety of dark, dense, fine-grained hornfels—occur locally. In the Del Mar quadrangle, fossil-bearing marine sedimentary rocks are interbedded with the volcanic and volcaniclastic rocks. Included with the Santiago Peak Volcanics are a number of small mildly metamorphosed gabbroic to granodioritic plutons which are considered to have been feeders for the volcanics rather than parts of the southern California batholith.

The Santiago Peak Volcanics, which form elevated peaks immediately east of the area at Black Mountain, are hard and extremely resistant to weathering and erosion. Most of the volcanic rocks are dark greenish gray where fresh and weather grayish red to dark reddish brown. The soil developed on the Santiago Peak Volcanics is the color of the weathered rock and supports the growth of dense chaparral.

Age estimates for the Santiago Peak Volcanics have ranged from Late Triassic (Hanna, 1926) to mid-Cretaceous (Milow and Ennis, 1961). Fife *et al.* (1967) reported latest Jurassic (Portlandian) fossils from a marine clastic part of the succession near Del Mar, and to date this constitutes the most reliable age for these rocks in the San Diego area.

Gabbro of the Southern California Batholith

Larsen (1948) named the batholithic rocks in the San Diego coastal area the Woodson Mountain Granodiorite, the Bonsall Tonalite, and the San Marcos Gabbro. Though most of the plutonic rocks in proximity to San Diego are quartz diorite and granodiorite, only gabbro crops out within this area. The gabbro varies considerably in texture and composition but generally is coarse grained and dark gray. The chief mineral constituents are calcic feldspar and pyroxene with minor amounts of quartz and biotite.

Potassium-argon dates (Evernden and Kistler, 1970) from a gabbro located 20 km northeast of Del Mar near San Marcos, and a quartz diorite located 10 km southeast of Escondido are, respectively, 101 and 105 million years. A lead-alpha date on zircon from quartz diorite in the Woodson Mountain area, 20 km southeast of Escondido, is 105 ± 10 million years (Bushee *et al.* 1963).

Throughout most of its exposure, the gabbro is weathered and difficult to distinguish from overlying sedimentary formations, which are largely composed of weathered plutonic basement rock. Careful examination for relict features, such as small quartz veins, is necessary to distinguish the weathered rock from the overlying sedimentary strata.

Rosario Group

The Upper Cretaceous Rosario Group is composed of clastic sedimentary rocks of marine and nonmarine origin assigned to the Lusardi, Point Loma, and Cabrillo Formations.

Lusardi Formation

The basal formation of the Rosario Group, the Lusardi Formation, was named by Nordstrom (1970) for exposures of boulder conglomerate near the confluence of Lusardi Creek and the San Dieguito River, 2 km north of the area in the Rancho Santa Fe quadrangle.

These rocks consist of cobble and boulder conglomerate, with occasional thin lenses of medium-grained sandstone. Some of the clasts are 10 m in diameter. The Lusardi Formation at the exposures within the northeast quarter of the Del Mar quadrangle, and within the type area to the north, has a maximum thickness of 125 meters. At the Holderness No. 1 well, 17 km southeast of the tip of the Point Loma Peninsula, rocks considered to belong to the Lusardi Formation are 82 m thick, whereas at the Point Loma No. 1 well, 10 km north of Point Loma, these rocks are 295-376 m thick (Hertlein and Grant, 1944, p. 38). The Lusardi Formation at its type area is unconformably overlain by Eocene rocks, but 16 km to the north near Carlsbad, it is overlain conformably by siltstone and sandstone of the Point Loma Formation (Kennedy and Moore, 1971b).

The Lusardi Formation is considered to be Late Cretaceous in age because it contains quartz diorite boulders eroded from the mid-Cretaceous southern California batholith, which has a minimum age of 105 ± 10 million years (Bushee *et al.*, 1963), and it is overlain by the Point Loma Formation which contains Upper Cretaceous (Campanian) Foraminifera (Sliter, 1968).

The Lusardi Formation is lithologically equivalent to the Trabuco Formation of the Santa Ana Mountains on the north (Nordstrom, 1970), to an unnamed fanglomerate near the base of the Williams Formation, also in the Santa Ana Mountains (Morton, 1972, p. 39), and to the Redondo Formation of Flynn (1970) in northern Baja California.

Point Loma Formation

The Point Loma Formation, the intermediate part of the Rosario Group, crops out along the sea cliffs on the west side of the Point Loma Peninsula, and in the La Jolla sea cliffs from Bird Rock to La Jolla Shores Beach (plates 2A,3A). At its type locality at the tip of Point Loma, it has an exposed thickness of 83 meters. The rocks there are interbedded fine-grained dusky-yellow sandstone and olive-gray clay shale that occur in graded beds about 30 centimeters (cm) thick.

Scuba-diving observations 1860 m offshore from the type locality show that ledgy pavement-like sandstone, similar to that in the lower half of the exposed section, continues to a depth of at least 37 m (Turner *et al.*, 1968, p. 8). With a shoreline dip of 6° E., it is postulated that 190 m of section may be added below low-tide level to the observed thickness of the formation. This submarine information, combined with interpolation from well logs, suggests that the total thickness of the Point Loma Formation at its type locality is about 300 m (section D-D').

Fossil Foraminifera and calcareous nannoplankton indicate a Late Cretaceous age for the Point Loma Formation (Sliter, 1968; Bukry and Kennedy, 1969). Foraminifera from near the base of the formation at Carlsbad are middle to upper Campanian in age, whereas those from the uppermost beds are lower Maestrichtian in age (Sliter, 1968).

The exposed part of the Point Loma Formation correlates with the Williams Formation and the upper part of the Ladd Formation in the Santa Ana Mountains (Popenoe *et al.*, 1960) and with the middle part of Beal's (1924) Formacion Rosario in northern Baja California.

Cabrillo Formation

The Cabrillo Formation, the uppermost unit of the Rosario Group, is exposed on the Point Loma Peninsula from the southern tip north to Sunset Cliffs. At Pacific Beach in the sea cliffs, it is exposed from 300 m south of False Point to Bird Rock on the north and at La Jolla in an S-shaped belt around the noses of the Pacific Beach syncline and Mount Soledad anticline. In the sea cliff at its type section 250 m east of the new Point Loma lighthouse, it consists of massive medium-grained sandstone and cross-bedded cobble conglomerate containing fresh plutonic and metavolcanic clasts but lacking red porphyritic rhyolite-tuff cobbles characteristic of nearby Eocene rocks.

Throughout the mapped area, the Cabrillo Formation conformably overlies the Point Loma Formation. The formation is 81 m thick at its type locality, where it is unconformably overlain by Pleistocene deposits. Along the sea cliff at False Point, it has a thickness of 170 meters.

A clam from the east flank of Mount Soledad within the lower 5 m of the Cabrillo Formation has been identified as *"Pharella" alta* (Gabb) and assigned to the Maestrichtian (L. Saul, written communication, 1969). The Cabrillo Formation correlates with the upper part of the Formacion Rosario of Beal (1924) in northern Baja California and possibly with the upper part of the Williams Formation in the Santa Ana Mountains.

EOCENE DEPOSITS

La Jolla Group

The La Jolla Group (La Jolla Formation of Hanna, 1926) ranges from moderately deep-water, fine-grained siltstone, to sandy beach and lagoonal facies, and coarse-grained continental sandstone and conglomerate. Deep water fine-grained facies predominate to the southwest, whereas the lagoonal and continental facies are more abundant to the northeast. These units include six partly intertonguing and partially time equivalent formations, which from oldest to youngest, are the Mount Soledad Formation, Delmar Formation, Torrey Sandstone, Ardath Shale, Scripps Formation, and Friars Formation (figure 4).

Mount Soledad Formation

Southwest of the Rose Canyon fault, which in Rose Canyon displaces rocks on its southwest side (figure 4; plate 2A; section B-B′), an Eocene marine cobble conglomerate and sandstone unit, designated the Mount Soledad Conglomerate (part of the Rose Canyon Shale Member of Hanna, 1926), rests unconformably on Upper Cretaceous rocks of the Cabrillo Formation. This formation crops out around the Mount Soledad anticline in La Jolla and northern Pacific Beach and south of Mission Bay on the southwest flank of the Pacific Beach syncline (plates 2A, 3A). Block diagrams 3, 5, and 6 (figure 4) show the stratigraphic relationship of the Mount Soledad Formation and the overlying and partly intertonguing Ardath Shale, Delmar Formation, and Torrey Sandstone. At its type locality on Mount Soledad, the formation is 69 m thick and consists of cobble conglomerate with minor beds of sandstone. The conglomerate content of the formation is variable to the southeast where it is locally composed entirely of medium-grained sandstone. The conglomerate commonly overlies similar Upper Cretaceous conglomerate of the Cabrillo Formation. The presence of distinctive red porphyritic, soda rhyolite-tuff clasts in the Mount Soledad Formation differentiates it from the Cabrillo Formation. This difference is easily seen at a sea-cliff exposure 300 m northwest of the end of Tourmaline Street in Pacific Beach where the two conglomerates are in contact (photo 2). The sandstone is moderately well sorted, subangular to subrounded, poorly indurated, and well bedded. It consists of quartz (75-80 percent), potassium feldspar (20-25 percent), plagioclase (1-2 percent), biotite (1-2 percent), and a trace of epidote, pyroxene, and hematite.

Photo 2. Unconformity between the Eocene and Upper Cretaceous rocks located 300 meters north of Tourmaline Street in Pacific Beach, looking northeast. The Eocene rocks contain soda rhyolite-tuff clasts not found in the Upper Cretaceous conglomerates.

The Ardath Shale, conformably overlying the Mount Soledad Formation, contains fossils which are lower middle Eocene in age (Bukry and Kennedy, 1969). The Mount Soledad Formation correlates with the basal part of the Santiago Formation in the Santa Ana Mountains (Kennedy and Moore, 1971a).

Delmar Formation

The Delmar Formation (Delmar Sand Member of Hanna, 1926) is exposed from the northern edge of the area mapped for 9 km south to Soledad Valley where it is overlain by younger rocks (plate 1A). The stratigraphic relationship of the Delmar Formation with the Mount Soledad Formation, Torrey Sandstone, and Ardath Shale is shown in figure 4.

Most of the Delmar Formation is dusky yellowish-green sandy claystone interbedded with medium-gray coarse-grained sandstone. Several resistant beds composed of *Ostrea idriaensis* Gabb and other brackish-water mollusks indicate a lagoonal origin. The sandstone is typically composed of quartz (80-85 percent), potassium feldspar (10-15 percent), plagioclase (1-2 percent), biotite (2-3 percent), and a trace of hematite, topaz, glauconite, and pyroxene. The claystone is composed of montmorillonite and kaolinite.

The base of the formation is not exposed but is presumed to rest unconformably on Upper Cretaceous or older rocks (section A-A′) or conformably on the Mount Soledad Formation as do correlative formations to the north and south (section B-B′). In its type section near the town of Del Mar and throughout the area, it is overlain gradationally by the Torrey Sandstone, with which it is also partly equivalent (figure 4). In the subsurface 15 km north near Carlsbad, the Delmar Formation grades into the Santiago Formation, and its boundary with the Santiago Formation occurs directly below the northernmost depositional limit of the overlying Torrey Sandstone (Kennedy and Moore, 1971a).

The Delmar Formation is considered to be middle Eocene in age because it is correlative in part with the Mount Soledad Formation on the south, the Santiago Formation on the north, and contains a rich Domengine molluscan assemblage.

Torrey Sandstone

The Torrey Sandstone (Torrey Sand Member of Hanna, 1926) crops out continuously from the northern boundary of the area 12 km south to Torrey Pines Golf Course and inland about 10 km (plates 1A, 2A). It has a maximum thickness of 60m and is composed of arkosic sandstone which is white to light brown, medium to coarse grained, subangular, and moderately well indurated. It is massive and broadly cross-bedded. The sandstone consists of quartz (85-90 percent), orthoclase (5-10 percent), plagioclase (less than 1 percent), biotite (1-5 percent), and a trace of hematite, epidote, zircon, tourmaline, pyroxene, and amphibole. At the type section at

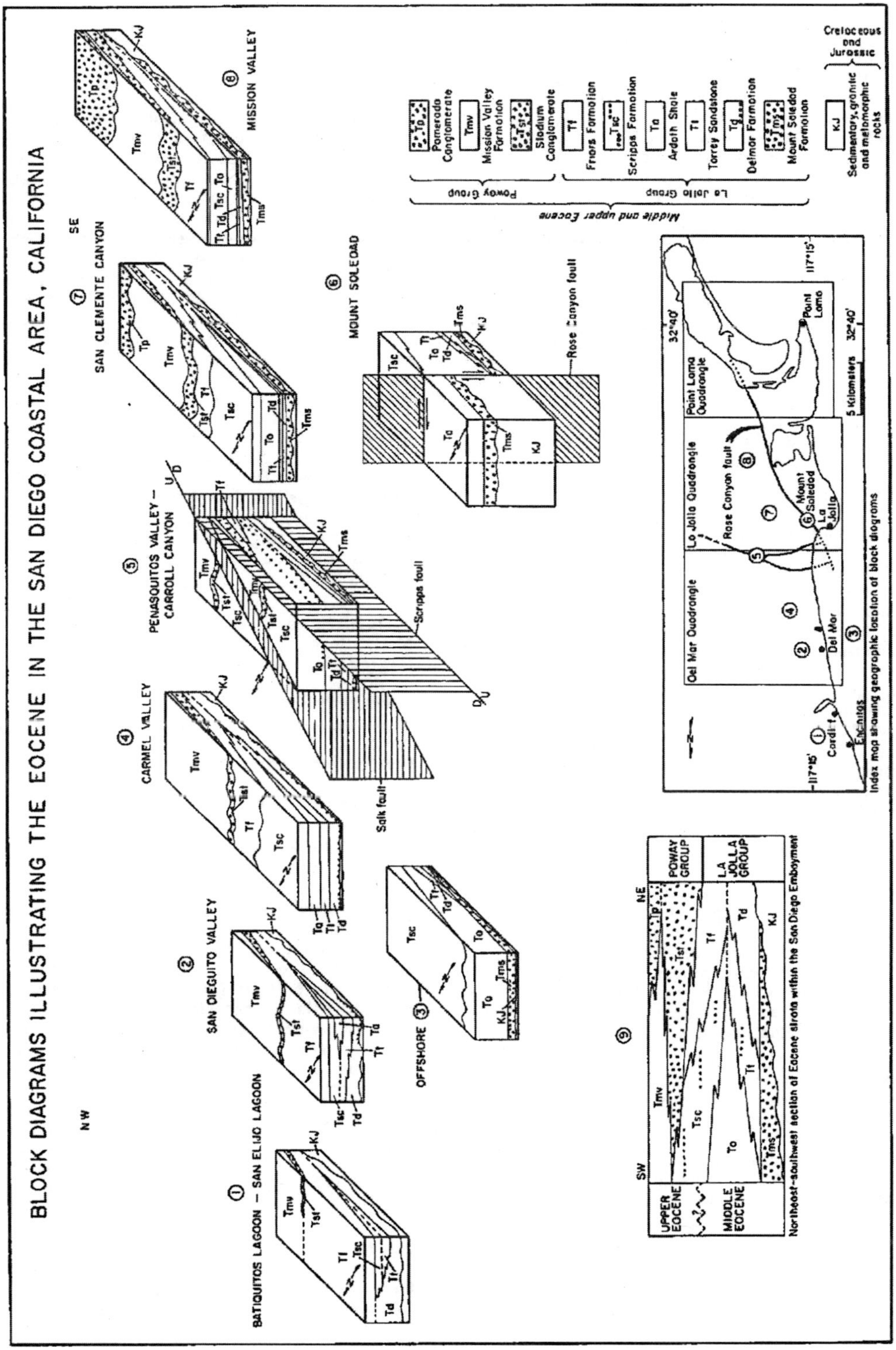

Figure 4. Block diagrams of the interrelationship between Eocene facies in the San Diego coastal area.

Torrey Pines grade on Highway 101, the contact with the underlying Delmar Formation consists of an alternating gradation between white sandstone beds above the dusky yellow-green fossiliferous claystone beds below. Approximately 15 km to the north, the Torrey Sandstone grades into and is overlain by the Santiago Formation. In Soledad Valley the lower part grades into and is overlain by the Ardath Shale and the upper part by the Scripps Formation (Kennedy and Moore, 1971a). The distribution and stratigraphic relationship of the Torrey Sandstone with respect to its related facies is shown in figure 4.

The Torrey Sandstone is believed to have been deposited along a submerging coast on an arcuate barrier beach that enclosed and then later transgressed over Delmar lagoonal sediments. Its deposition was arrested when submergence slowed and the shoreline retreated. Although the Torrey Sandstone contains only a few poorly preserved fossils and fossil casts, its middle Eocene age is firmly established by its interfingering relationship with the well-dated Ardath Shale.

Ardath Shale

The Ardath Shale (part of the Rose Canyon Shale Member of Hanna, 1926) crops out along the sea cliffs from Bathtub Rock in Torrey Pines State Park, south to the pier at Scripps Institution of Oceanography, where it is overlain by Pleistocene deposits(plates 1A, 2A). From Rose Canyon it can be traced south to the northeast corner of Mission Bay, where it is overlain by younger rocks. In the northwest corner of the area, it thins below the Scripps Formation as it grades into the Torrey Sandstone below. The stratigraphic relationship between the Ardath Shale and formations with which it interfingers is illustrated in figure 4. The Ardath Shale is predominantly weakly fissile, olive-gray shale. Concretionary beds containing molluscan fossils are common. Expansible claystone locally comprises as much as 25 percent of the unit and landslides are commonly associated with those areas. Sieve analyses indicate that the particle size distribution is 81 percent silt, 16 percent clay, and 3 percent sand. The clay is mostly kaolinite but montmorillonite is also present. The sand consists of quartz (70-75 percent), potassium feldspar (15-20 percent), biotite (5-10 percent), plagioclase (less than 1 percent), and a trace of zircon, tourmaline, pyroxene, and amphibole.

The base of the Ardath Shale is not exposed at its type section in Rose Canyon (Kennedy and Moore, 1971a), but underlying outcrops at the type locality of the Mount Soledad Formation, 1 km to the west, reveal the contact on the Mount Soledad Formation to be conformable. The Ardath Shale is estimated to be 70 m thick at its type locality. It grades alternatingly and conformably into the overlying Scripps Formation.

Abundant fossils, including mollusks and calcareous nannoplankton, permit an assignment of the Ardath Shale to the lower middle Eocene (Bukry and Kennedy, 1969). The unit correlates with the basal part of the Torrey Sandstone and with the middle part of the Santiago Formation.

Scripps Formation

The Scripps Formation (part of the Rose Canyon Shale Member of Hanna, 1926) underlies much of the area from east of Del Mar on the north to the mouth of Mission Valley on the south (plates 1A-3A). Along the coast, it extends from the middle of Torrey Pines Park to Scripps Institution of Oceanography. The type section of the Scripps Formation is 1 km north of Scripps Pier, on the north side of the mouth of Blacks Canyon (Kennedy and Moore, 1971a). Here it consists of 56 m of pale yellowish-brown, medium-grained sandstone and occasional cobble-conglomerate interbeds. The sandstone is composed of subangular grains of quartz (75-80 percent), potassium feldspar (15-20 percent), biotite (2-5 percent), plagioclase (less than 1 percent), and a trace of epidote, pyroxene, tourmaline, sphene, and apatite. Both the basal contact with the Ardath Shale and the upper contact with the Friars Formation are conformable. Figure 4 illustrates the stratigraphic relationship between the Scripps Formation and related facies. Several tongues belonging to the Scripps Formation occur in the section. The largest of these is mapped in the vicinity of Sorrento Valley (plate 1A).

Fossils are present but are less common in the Scripps Formation than in the underlying Ardath Shale. Because of its close conformity and partial intergradation with the Ardath Shale, the Scripps Formation is considered to be middle Eocene. To the north near Encinitas it correlates with the upper part of the Torrey Sandstone and, farther to the north in the Santa Ana Mountains, with the middle part of the Santiago Formation.

Friars Formation

The middle and late Eocene Friars Formation (part of the Rose Canyon Shale Member of Hanna, 1926) is the uppermost unit of the La Jolla Group. The rocks are nonmarine and lagoonal sandstone and claystone named for exposures along the north side of Mission Valley near Friars Road in the La Jolla quadrangle (Kennedy and Moore, 1971a). The sandstone is composed of quartz (75-80 percent), potassium feldspar (10-15 percent), biotite (5-10 percent), plagioclase (less than 1 percent), and a trace of amphibole, pyroxene, hematite, and tourmaline. The claystone is composed of montmorillonite and kaolinite. Friars Formation is predominantly a nonmarine and nearshore marine facies which reaches a maximum thickness of 50 m between Mission Valley and Carmel Valley. The sandstone is typically massive, yellowish gray, medium grained, and poorly indurated with subangular to subrounded grains. Caliche-rich sandstone beds are locally interlayered with dark greenish-gray sandy claystone. Cobble conglomerate lenses and tongues of fluviatile origin are characteristic of the easternmost exposures. The Friars Formation rests unconformably on rocks of the basement complex and conformably on the Scripps

Formation. It is in turn overlain by other sedimentary deposits of Eocene, Pleistocene, and Holocene age. The Friars and Delmar Formations are lithologically identical in their central and northeastern exposures, and they have been undifferentiated in these areas on the geologic map.

The stratigraphic relationship between the Friars Formation and related facies is diagrammatically illustrated in figure 4.

Poway Group

The Poway Group (Poway Conglomerate of Hanna, 1926) includes three partly intertonguing and partially time equivalent formations, the Stadium Conglomerate, the Mission Valley Formation, and the Pomerado Conglomerate. These rocks are primarily nonmarine in their easternmost exposures and nearshore marine and lagoonal in their westernmost exposures.

Stadium Conglomerate

The type section of the Stadium Conglomerate lies near the boundary between the La Jolla and La Mesa quadrangles along the northern wall of Mission Valley near San Diego Stadium (Kennedy and Moore, 1971a). At the type section the formation consists of a massive cobble conglomerate with a dark yellowish-brown coarse-grained sandstone matrix. The conglomerate contains dispersed lenses of fossiliferous crossbedded sandstone. The fossils include calcareous nannoplankton of late? Eocene age.

The Stadium Conglomerate is moderately well sorted with an average clast size in the cobble range. Clasts having diameters as large as .05 m do occur but are rare. The sandstone matrix constitutes less than 20 percent of the unit, but in local stratigraphic sections individual sandstone beds and lenses constitute 50 percent of the unit.

The highly distinctive "Poway" clasts that occur only within Cenozoic deposits of southern California and that typify the Stadium Conglomerate consist predominantly (up to 85 percent) of slightly metamorphosed rhyolitic to dacitic volcanic and volcaniclastic rocks and up to 20 percent quartzite. The source area for these clasts is controversial, and potential sources from the Mojave Desert to Sonora, Mexico, have been proposed (DeLisle *et al.*, 1965; Merriam, 1968; Woodford *et al.*, 1968; Minch, 1972). Based on direction of pinching and cobble imbrication, the clasts within the Stadium Conglomerate were deposited within the San Diego embayment by a westward-flowing river system. Based on clast size the conglomerates were probably derived from a now eroded source within 150 km of their present position (Kennedy, 1973a).

The Stadium Conglomerate is conformably underlain by the Friars Formation and is conformably overlain by the Mission Valley Formation. The stratigraphic relationship between the Stadium Conglomerate and genetically related Eocene conglomerate to the east is shown in figure 4.

Mission Valley Formation

The Mission Valley Formation, a predominantly marine sandstone unit, lies conformably upon the Stadium Conglomerate and is in turn conformably overlain by the Pomerado Conglomerate. It has a maximum thickness of 60 m and was named for exposures along the south wall of Mission Valley on the west side of State Highway 163 (Kennedy and Moore, 1971a). The sandstone is characteristically soft and friable, light olive gray, fine to medium grained, and composed mostly of quartz and potassium feldspar. The grains are subangular to subrounded and locally range in size from coarse to very fine sand. Plagioclase and biotite are also present but generally constitute less than 2 percent each. Cobble conglomerate tongues within the Mission Valley Formation, similar to Stadium Conglomerate, comprise up to 30 percent of sections measured in the easternmost exposures of the area but less than 1 percent of sections measured along the westernmost outcrops.

Because of the friable nature of the Mission Valley Formation, it lacks the bold topographic expression displayed by the Stadium Conglomerate. Interbeds and tongues of claystone of brackish water origin locally compose 20 percent of the section. The clay is primarily montmorillonite but kaolinite is also present. The Mission Valley Formation thins from the west to the east (figure 4) and pinches out in the eastern part of the Poway and La Mesa quadrangles. The rocks are often fossiliferous and contain a molluscan fauna in the western and central exposures and a land-mammal fauna in the eastern exposures. One molluscan assemblage, collected from the uppermost beds of the formation in a road cut 200 m due east of the Miramar Reservoir filtration plant (elevation 238 m) at Lat. 32° 54.8' N.; Long. 117° 05.7' W., is reported by C.R. Givens (written communication, 1970) to be characteristic of the upper Eocene (Tejon Stage) and correlative with the upper Eocene of Europe.

Pomerado Conglomerate

The Pomerado Conglomerate, the uppermost formation of the Poway Group, has a maximum thickness of 55 meters. It was named for exposures located at the divide between Carroll Canyon and Poway Valley along Pomerado Road east of the area in the Poway quadrangle (Peterson and Kennedy, 1974). The Pomerado Conglomerate is late Eocene in age and is a massive cobble conglomerate lithologically identical to the Stadium Conglomerate. The contact between the Mission Valley Formation and Pomerado Conglomerate is conformable and gradational. The Pomerado Conglomerate is characterized by occasional thin beds, lenses, and tongues of light-brown medium-grained sandstone. One of the largest of these, which crops out east of the area near Miramar Reservoir in the Poway quadrangle, is designated the Miramar Sandstone Member (Peterson and Kennedy, 1974). Lithologically the Miramar Sandstone Member is identical to the Mission Valley Formation but is stratigraphically higher and wholly contained within the Pomerado Conglomerate. It has a maximum thickness of 10 m in its type area and is considered to be late Eocene in age based on its superpositional relationship with the Pomerado Conglomerate and Mission Valley Formations.

FACIES RELATIONSHIPS OF THE EOCENE ROCKS

The Eocene rocks of the San Diego embayment were laid down on a narrow continental shelf and adjacent margin that extended northwest and southeast for more than 50 kilometers. Subsidence of the basin and repeated change in sediment flux resulted in alternating advances and retreats of the shoreline. The advances are recorded by the deposition of time-transgressive rock units and the retreats by their time-regressive counterparts (Kennedy, 1971; 1973).

Nonmarine lagoonal and nearshore marine facies were deposited on the east and marine continental shelf facies on the west side of the San Diego embayment. There are two lithostratigraphic groups divided into nine intertonguing formations that together are approximately 700 m thick. The formational names are those of Kennedy and Moore (1971a) and Peterson and Kennedy (1974).

The two groups are the La Jolla and Poway. The La Jolla Group is stratigraphically lower and lies predominantly west of the Poway Group. Deltaic conglomerate and sandstone, lagoonal sandstone and claystone, beach sandstone, and marine shale constitute the La Jolla Group. Deltaic conglomerate and sandstone, lagoonal sandstone, and littoral sandstone and siltstone comprise the Poway Group. Figure 4 illustrates the interrelationships of the rocks and the contact between the groups.

Deposition occurred continuously in the San Diego embayment for a period of nearly 10 million years during which time the regional tectonic downwarping of the basin took place. The subsidence is marked in the stratigraphic record by two prominent marine transgressions and two intervening regressions.

There are two somewhat conflicting hypotheses used to explain the development of cyclic sedimentation of this type (Sears *et al.*, 1941). The cyclic stratigraphic succession that forms by either of the two is an intertonguing sequence of strata having time regressive and transgressive parts (i.e., marine, littoral, beach, nonmarine) that grade laterally and vertically with respect to each other.

One hypothesis or model, which has been rejected for the development of the San Diego embayment, involves alternating upward and downward movement of the marine basin and adjacent continental land mass to create the necessary change in sea level. The erosion of previously deposited materials by waves, during times when uplift occurred faster than sedimentation, would have removed parts of the cyclic facies. The cycles in the San Diego embayment are gradational, complete, and considered to have originated under different conditions.

Another model, and the one suggested for the development of the Eocene facies here, is based upon continuous subsidence of the sedimentary basin with changes occurring in both the rate of subsidence and rate of sedimentation. Regressive deposits are formed by the slowing of subsidence and/or an increase in sedimentation which creates the outward building of the shoreline and shallowing by infilling of the basin. During periods of more rapid subsidence and/or the slowing of sedimentation rate, the basin deepens and transgressive deposits are laid down.

Figure 5 is a diagrammatic illustration of the second model. Beginning at the top of figure 5 with diagram A, subsidence is occurring at a rapid rate with respect to sediment influx. This shows the initial development of the lagoon, beach, and nearshore deposits. Diagrams B and C illustrate later time but with continued conditions as represented by A. Note that these units are time-transgressive and that sea level has advanced over the old land surface. Beginning with diagram D either a slowing in the rate of subsidence has occurred or the sedimentation rate has increased or both. As a result of this change, the first regressive deposits are formed and the shoreline retreats.

A period of rapid subsidence and high sedimentation marks the beginning of the first transgression, recorded by the deposition of the Delmar Formation, Torrey Sandstone, and Ardath Shale. These sediments transgressed eastwardly over and beyond the Mount Soledad Formation into pre-Eocene basement rock (section B-B'). The Ardath Shale rests conformably upon the Mount Soledad Formation at the type section of the Mount Soledad Formation located 600 m east of Easter Cross in La Jolla. The Torrey Sandstone rests gradationally upon the Mount Soledad Formation at the base of Indian Trail in the sea cliffs 3300 m north of Scripps pier and at the intersection of Carmel Valley and Soledad Valley (plate 1A). The conglomerate shown within the lower Delmar Formation is considered to represent the transitional facies between the Delmar and Mount Soledad Formations.

As shown in figure 4 and plates 1A-3A, the transgressive nature of these stratigraphic units is indicated by their superpositional and lateral relationship. The lagoonal deposits are predominantly low in the section and lie to the east and northeast of the beach-bar and marine deposits. The beach and beach-bar deposits grade laterally eastward and downward into lagoonal deposits and westward and upward into marine deposits. Marine deposits are high in the section and lie to the west of the beach and lagoonal deposits.

A slowing in subsidence and/or an increase in sedimentation to a degree that allowed infilling of the embayment at a greater rate than subsidence marks the beginning of a retreating shoreline and the development of regressive deposition.

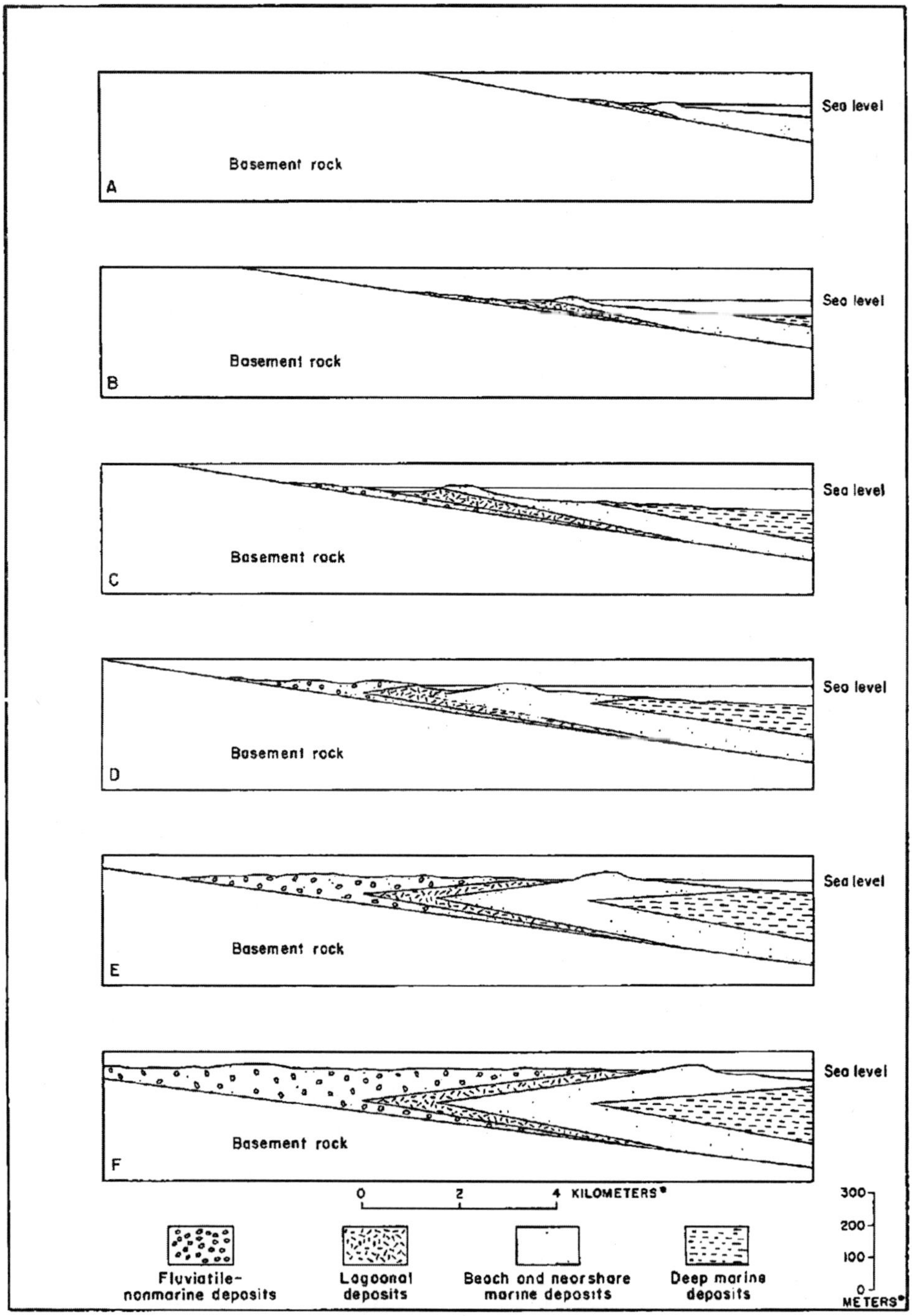

Figure 5. Model of transgressive and regressive deposition. (Modified from Sears *et al.*, 1941).

As shallowing occurred the lagoonal, beach, and marine deposits migrated westward creating a reversal in their superpositional relationship (figure 4). Again the lagoonal rocks lie predominantly on the east and marine rocks predominantly on the west, but the regressive lagoonal rocks comprise the upper part of the stratigraphic sequence and the marine rocks the lower part.

The regressive lagoonal equivalent of the transgressive Delmar Formation is the Friars Formation and that of the beach-bar Torrey Sandstone is the Scripps Formation. The Ardath Shale also has a trangressive and regressive phase; however, these have not been separated, as they constitute a continuous section that is lithologically homogeneous.

Interbeds, tongues, and lenses of cobble conglomerate composed of exotic tuffaceous clasts, mostly of rhyolitic composition, and the primary sediment of a significant westwardly or northwestwardly flowing river system are abundant in this part of the section. The direction of transport is indicated primarily by cobble imbrication and paleostream channel mapping (Minch, 1972). In the central and eastern part of the embayment, thick deposits of these clasts form the Stadium and Pomerado Conglomerates.

Renewed subsidence and/or a slowing of deposition marked a second transgressive cycle and rocks of nearshore marine and nonmarine origin were laid down. The transgressive nature of the strata can be detected by the superpositional and gradational relationship between and within the Mission Valley Formation and the Stadium Conglomerate. The Mission Valley Formation is the continuum and regressive equivalent of the Scripps Formation (figure 4).

A final regressive conglomerate unit, the Pomerado Conglomerate (figure 4), has been preserved high in the stratigraphic succession. One short period transgression marked by the Miramar tongue near the Miramar Reservoir in the Poway quadrangle is the uppermost marine sandstone of the column.

EOCENE BIOSTRATIGRAPHY

The Eocene lithostratigraphic succession discussed in the preceding pages contains fossil organisms representative of deep water marine, littoral marine, lagoonal, and nonmarine fluviatile environments. These fossils together indicate that the boundary between the middle and late Eocene lies lies near the boundary between the La Jolla and Poway Groups in the central exposed part of the San Diego embayment and that the middle and late Eocene boundary falls within the Uintan Mammal Age (Golz and Kennedy, 1971). This is later, relative to the base of the Uintan, than originally proposed (Wood *et al.*, 1941). The Eocene succession of the San Diego embayment is presently the only place known in North America where this part of the Tertiary mammal chronology can be directly compared with invertebrate chronologies (figure 6).

Four major fossil groups have been collected from the Eocene rocks of the San Diego embayment. These include 1) mammals, 2) mollusks, 3) calcareous nannoplankton, and 4) Foraminifera (figures 7-9).

1) Mammalian fossils were collected from the Ardath Shale, Friars Formation, Stadium Conglomerate, and Mission Valley Formation. The collection has been studied (Golz, 1971, 1973) and found to be correlative in its stratigraphically uppermost part with the North American Uinta C Mammal Age and in its lowest part with Bridgerian Mammal Age.

2) Molluscan fossils were collected from Mount Soledad Formation, Delmar Formation, Torrey Sandstone, Ardath Shale, Scripps Formation, Friars Formation, and Mission Valley Formation. The fossils have been correlated, using West Coast (California) Molluscan Stages, with the Tejon Stage in the stratigraphically uppermost part of the section, the "Transition Stage" in the intermediate part, and the Domengine Stage in the lower part (Hanna, 1926; Moore, 1968; C. R. Givens, written communication, 1973).

3) Calcareous nannoplankton have been collected from the Ardath Shale, Scripps Formation, Stadium Conglomerate, and Mission Valley Formation. The flora in the stratigraphically lowest part of the section is indicative of the middle Eocene, Lutetian Stage of Europe (Bukry and Kennedy, 1969). The flora from the uppermost beds collected is sparse and is questionably correlative with the lower part of the upper Eocene (D. Bukry, written communication, 1971).

4) Foraminifera have been collected from the Ardath Shale, Scripps Formation, Friars Formation, and Stadium Conglomerate. The fauna from the stratigraphically uppermost part of the section has been reported by Mallory (1959) to be correlative with his Narizian Stage (late Eocene age) and that from the lower part of the section with his Ulatisian Stage (middle Eocene age). Steineck and Gibson (1971) have studied Foraminifera from both the Ardath Shale and Stadium Conglomerate and report a middle Eocene and late middle Eocene age respectively.

The discussion that follows establishes the fossil composition of each of the nine lithostratigraphic units within the Eocene San Diego embayment and relates the West Coast (California) molluscan stages to the North American Mammal Ages and these two chronologies to the Eocene of Europe by way of correlations based on planktonic calcareous nannoplankton zones.

A relatively rich molluscan assemblage has been collected from the middle part of the Mount Soledad Formation at localities 1 and 2 (figure 7; (plates 2A, 3A). These localities combined are represented by M1 in figure 6. The assemblage from these localities includes *Turritella uvasana applinae* Hanna, *Ficopsis cooperiana* Stewart, and *Tejonia lajollaensis* (Stewart), all of which are restricted to the middle Eocene Domengine Stage of California (Givens, 1974).

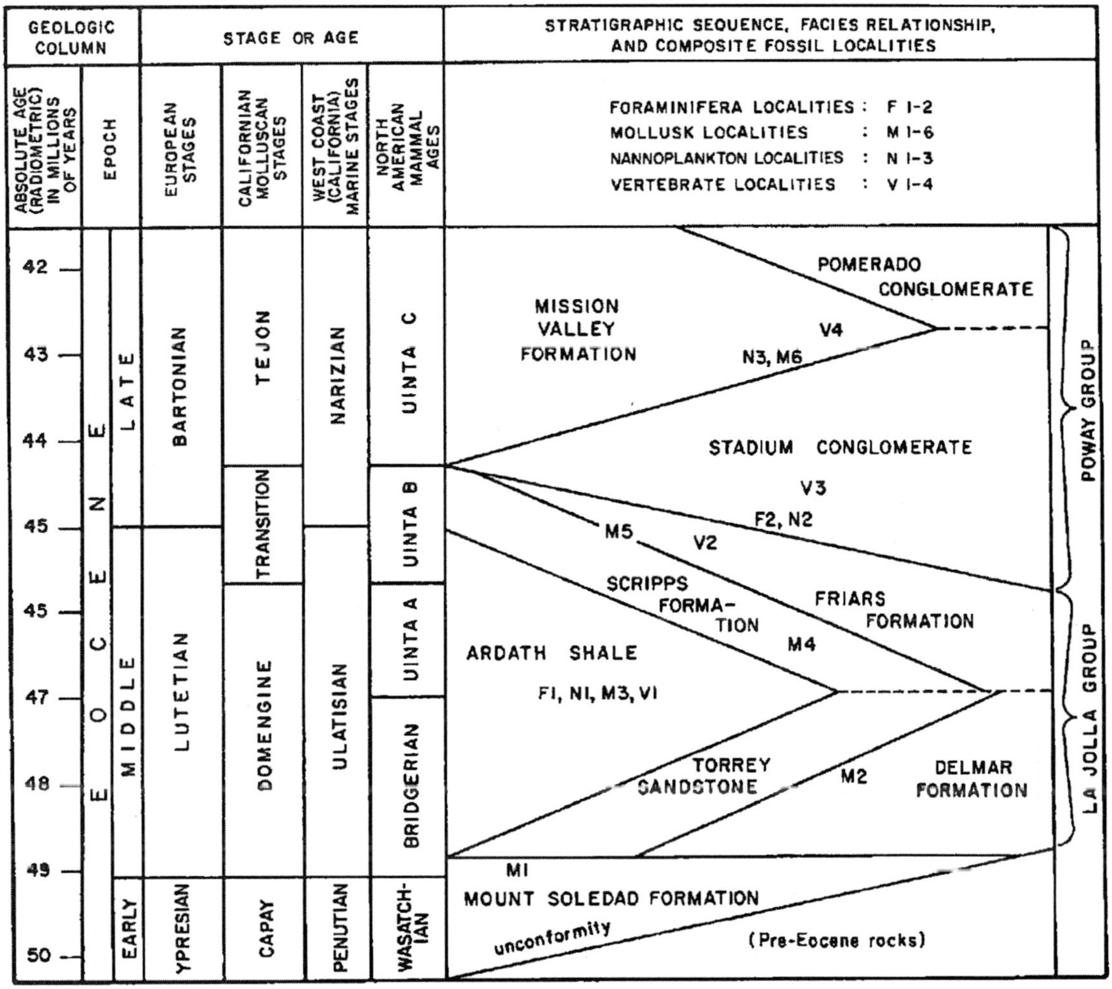

Figure 6. Relationship of biostratigraphy to lithostratigraphy.

The upper contact between the Mount Soledad Formation in its type section near La Jolla with the overlying Ardath Shale is conformable. The age of calcareous nannoplankton in the Ardath Shale at localities 9 and 11, less than 25 m above this contact, are middle Eocene in age (Bukry and Kennedy, 1969). The age assigned to the middle and upper parts of the Mount Soledad Formation is middle Eocene.

The Mount Soledad Formation interfingers with the Delmar Formation to the north where a rich molluscan fauna has been collected from rocks at localities 3, 4, and 5 (figure 7; plate 1A). These localities combined are represented by M2 in figure 6.

Horizons composed of nearly pure fossil shell material, mostly derived from brackish-water oysters, primarily *Ostrea idriaensis* Gabb, are abundant in the Delmar Formation. *Tejonia lajollaensis* (Stewart), restricted to the Domengine Stage of California (Givens, 1974), has been collected at localities 3 and 4.

Because the Delmar Formation is also chronologically correlative with parts of the Ardath Shale (figure 4), the Delmar Formation is assigned a middle Eocene age. Based on a similar age, stratigraphic position, lithology, and environment of deposition—the Delmar is considered to correlate with the lower part of the Santiago Formation in the Santa Ana Mountains to the north.

A sparse molluscan fauna was collected stratigraphically above the Delmar Formation from the Torrey Sandstone. The fauna consists of casts of *Ostrea idriaensis* Gabb and unidentifiable fragments of other pelecypods. A middle Eocene age is again assigned to the Torrey Sandstone on the basis of its interfingered relationship with the Delmar Formation and Ardath Shale.

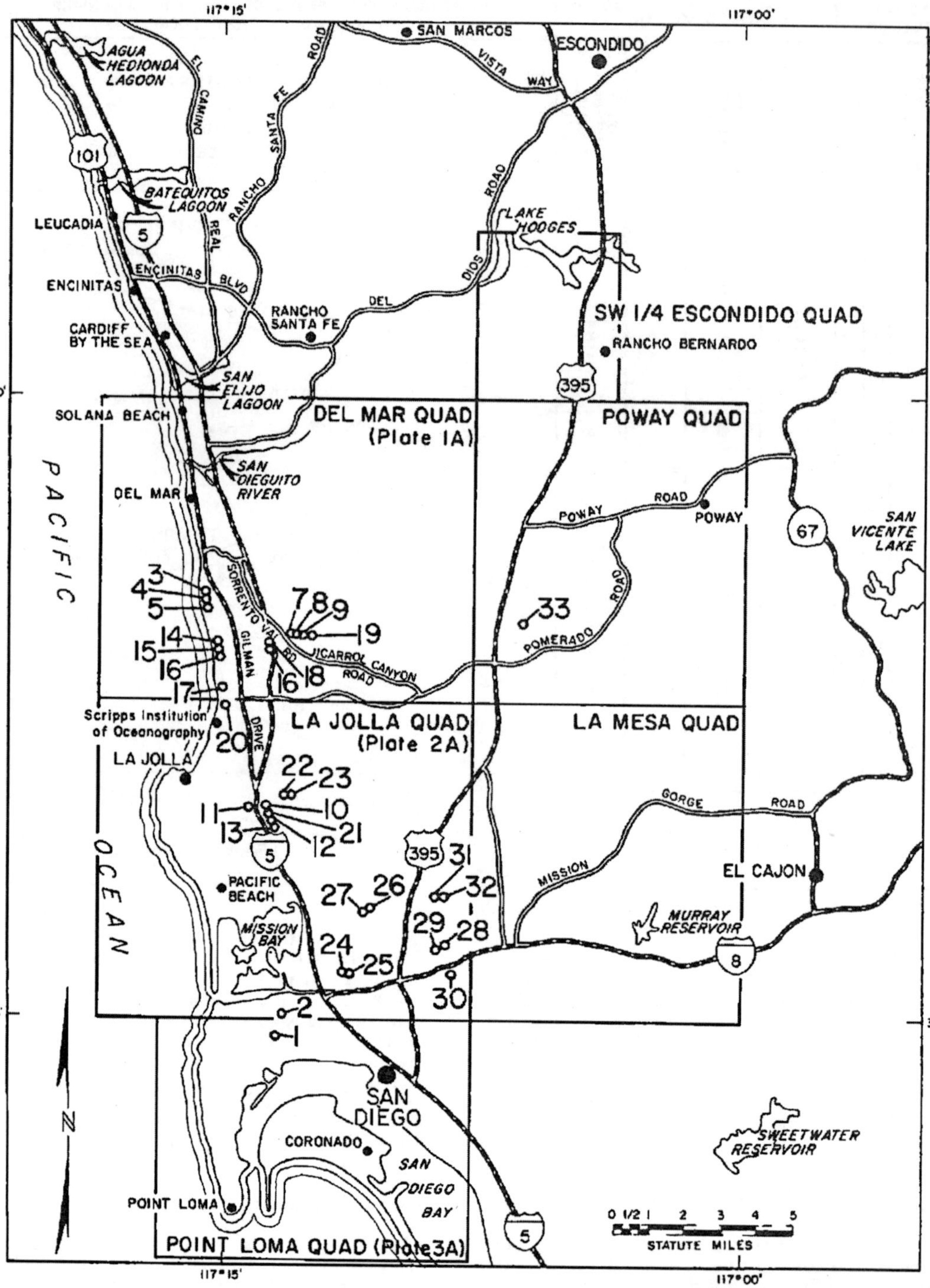

Figure 7. Index map of fossil mollusk localities. (See plates 1A-3A and plate 2B (Kennedy and Peterson, 1975) for detailed locations.)

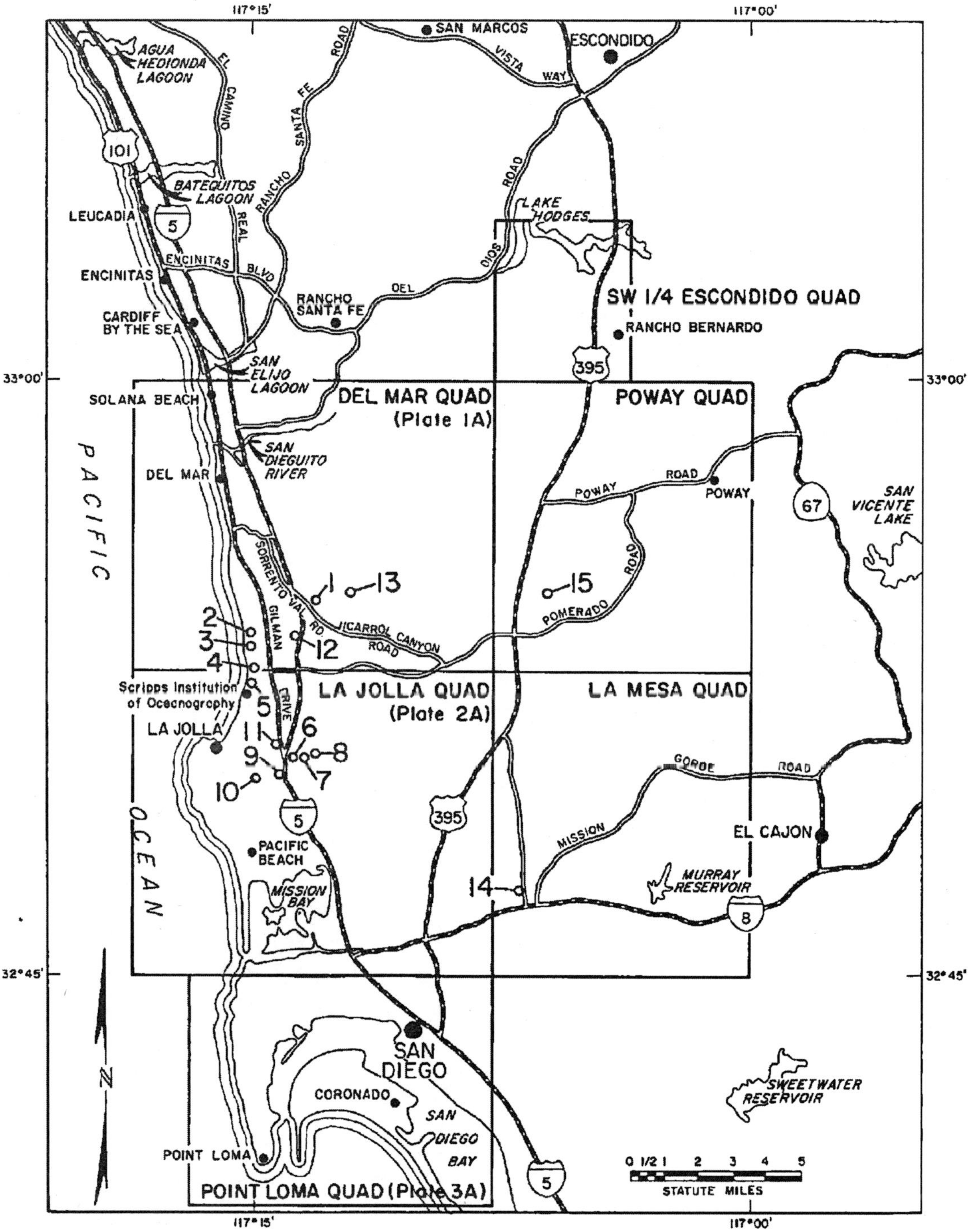

Figure 8. Index map of fossil calcareous nannoplankton localities. (See plates 1A, 2A and plate 2B, 3B (Kennedy and Peterson, 1975) for detailed locations.)

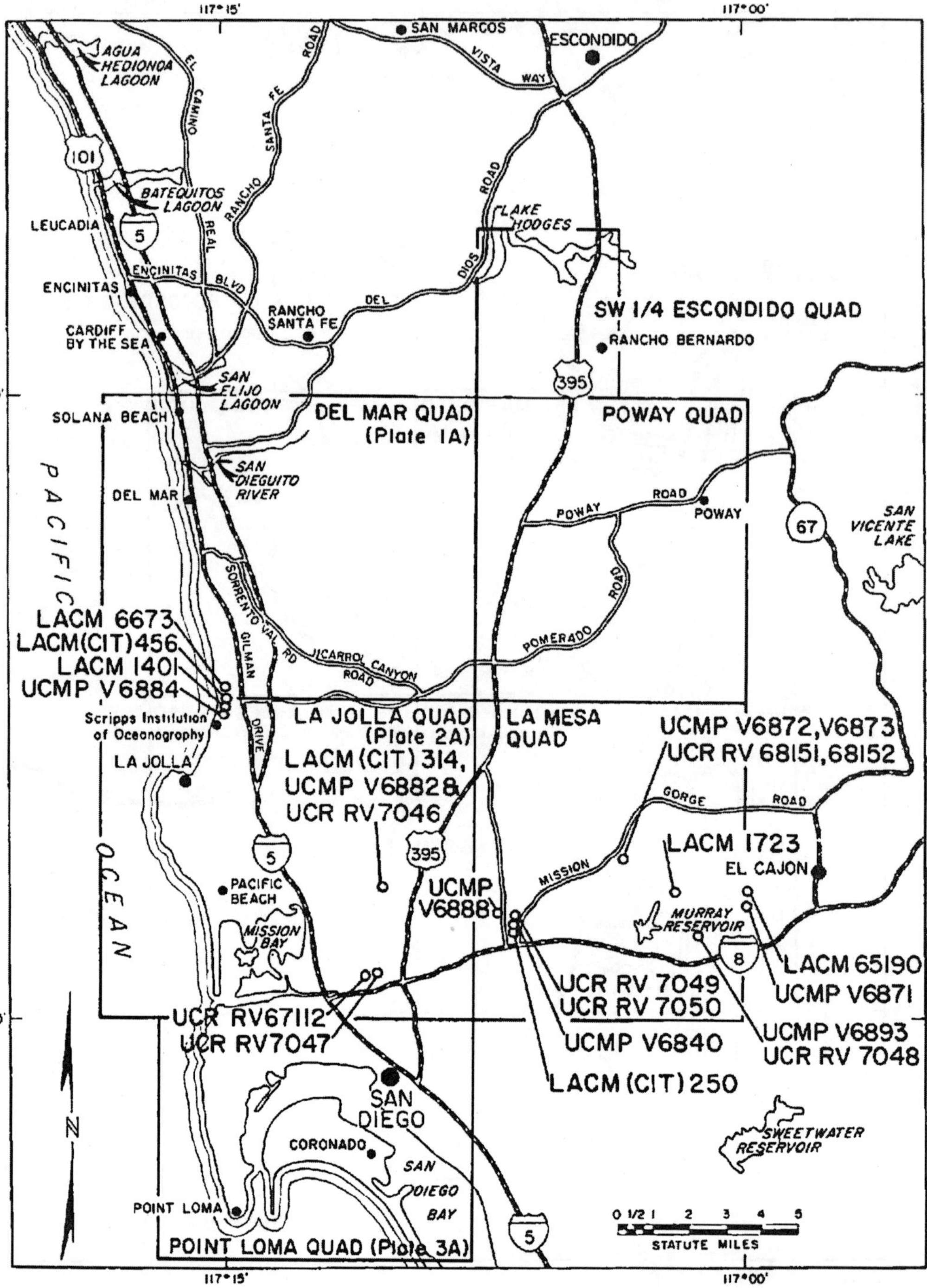

Figure 9. Index map of fossil mammal localities. (See plates 1A, 2A and plate 3B (Kennedy and Peterson, 1975) for detailed locations.)

Fossil mollusks, calcareous nannoplankton, mammals, and Foraminifera have been collected from the Ardath Shale. The molluscan fauna was collected from localities 6 through 16 (figure 7; plates 1A,2A). The localities combined are indicated by M3 in figure 6. The calcareous nannoplankton flora was collected from localities 1 through 11 (figure 8; plates 1A,2A). These localities combined are indicated by N1 in figure 6. Fossil mammals have been collected from Los Angeles County Museum localities LACM 6673, LACM (CIT) 456, and LACM 1401 and from the University of California, Berkeley, locality UCMP V6884 (figure 9; plates 1A, 2A). These localities are combined in figure 6 as V1. Foraminifera have been collected from the Ardath Shale at a locality 1.5 km north of its type section in Rose Canyon (Steineck and Gibson, 1971). This locality is indicated by F1 in figure 6.

The molluscan assemblage collected includes *Turritella uvasana applinae* Hanna, *Ficopsis cooperiana* Stewart, and *Tejonia lajollaensis* (Stewart), all of which are restricted to the Domengine Stage of California (Givens, 1974). Calcareous nannoplankton from localities 6 through 9 occur at the same stratigraphic interval as mollusks from localities 10 and 11. Nannoplankton locality 9 (figure 8; plate 2A), locality L13 of Bukry and Kennedy (1969), yields fossils indicative of an early middle Eocene age (*Discoaster sublodoensis* zone). These include *Coccolithus eopelagicus* (Bramlette and Riedel), *Coccolithus pelagicus* (Wallich), *Helicopontospaera seminulum lophota* (Bramlette and Sullivan), *Discoaster distinctus* Martini, and *Discoaster barbadiensis* Tan. Calcareous nannoplankton, also of early middle Eocene age (*Discoaster sublodoensis* zone), were collected from locality 11 (figure 8; plate 2A) in Rose Canyon from the type section of Hanna's (1926) Rose Canyon Shale Member. Steineck and Gibson (1971, p. 477) collected from this same locality and reported that the "occurrence of *Subbotina patagonica* Todd in both the Rose Canyon Shale and Cozy Dell Formation suggests time-equivalence of the two units (early middle Eocene in age)."

The fossil mammals collected from the Ardath Shale have been reported by Golz (1973) to be of Uinta A or Bridgerian Age. However, since these land animals were transported to a marine depositional environment, it is possible that they are older than the early middle Eocene rocks in which they lie.

The Ardath Shale is conformably and gradationally overlain by richly fossiliferous rocks of the Scripps Formation. Fossil mollusks were collected within the stratigraphically lower and intermediate part of the Scripps Formation from localities 17 through 23 (figure 7; plates 1A,2A) and are indicated together as M4 in figure 6. Mollusks were also collected from the upper part of the Scripps Formation near its contact with the Friars Formation at localities 24 and 25 (figure 7; plate 2A) and are indicated together as M5 in figure 6.

A molluscan assemblage from a conglomerate at the base of the Scripps Formation 400 m north of its type section at locality 17 (figure 7; plate 1A) includes *Turritella andersoni lawsoni* Dickerson, *Turritella uvasana applinae* Hanna, *Tejonia lajollaensis* (Stewart), and *Ficopsis cooperiana* Stewart. These fossils together indicate a Domengine Age (Givens, 1974). An assemblage from locality 25 (figure 7; plate 2A) 20 m below the contact between the Scripps Formation and the overlying Friars Formation about 400 m west of the type section of the Friars includes *Nekewis io* (Gabb) and *Ectinochilus canalifer supraplicatus* (Gabb). The upper part of the Scripps Formation based on the co-occurrence of these species is considered to belong to the "Transition Stage" and to be middle or late Eocene in age (C. R. Givens, written communication, 1973).

The Scripps Formation is in part laterally equivalent to and in part conformably overlain by rocks of the Friars Formation. Mollusks were collected within the Friars Formation from near the gradational boundary with the underlying Scripps Formation at localities 26 and 27 (figure 7; plate 2A). These localities, combined because of their close stratigraphic proximity to locality 24 and 25, also are indicated by M5 in figure 6. The mammal fauna was collected from seven localities in the Mission Valley-Mission Gorge region. These include the University of California, Riverside, localities UCR RV 7046, RV 67112, RV 7047, RV 7049, RV 7050, RV 68151, and RV 68152; University of California, Berkeley, localities UCMP V6872, V 6873, and V6888; and Los Angeles County Museum localities LACM (CIT) 250 and 314 (figure 9; Kennedy and Peterson, 1975, plate 3B). These localities together are indicated by V2 in figure 6.

The molluscan fauna collected includes *Nekewis io* (Gabb) and *Ectinochilus canalifer supraplicatus* (Gabb). These species together suggest that the lower part of the Friars Formation belongs to the "Transition Stage" and is middle and late Eocene in age (Givens, 1974).

The mammalian fauna collected from the Friars Formation has been studied by Golz (1973). He reports that the stage of evolution of the artiodactyl fauna from the Friars Formation in the upper Tecolote Creek and Mission Valley-Mission Gorge area is indicative of Uinta B Age.

Fossil calcareous nannoplankton, mammals, and planktonic Foraminifera have been collected from the Stadium Conglomerate. The nannoplankton were collected from a siltstone interbed at locality 14 (figure 8; plate 3B, Kennedy and Peterson, 1975) near the intersection of Murphy Canyon and Mission Valley. This locality is indicated as N2 in figure 6. The mammalian fauna was collected from Los Angeles County Museum locality LACM 1723 and from University of California, Berkeley, locality UCMP V6840 (figure 9; plate 3B, Kennedy and Peterson, 1975). Together these localities are indicated as V3 in figure 6. The Foraminifera were collected from nannoplankton locality 14 by Steineck and Gibson (1971). This locality is indicated as F2 in figure 6.

The nannofossils collected include *Reticulofenestra umbilica* (Levin) and *Discoaster distinctus* Martini. Because *Reticulofenestra umbilica* ranges from the upper middle Eocene to lower Oligocene, the age of these samples based on nannofossils is questionable.

The mammalian fauna, which occurs stratigraphically higher and to the east of the marine fauna and flora, has been correlated with the Uinta B or low Uinta C (Golz, 1973).

Foraminifera from the Murphy Canyon locality include *Globorotaloides suteri* Bolli and *Truncorotaloides collacteus* Finlay (Steineck and Gibson, 1971, p. 478). Though Steineck and Gibson state that this "co-occurrence suggests equivalence with upper middle Eocene strata", Jenkins (1965, figure 2) considers that the occurrence of these two species indicates a restricted upper Eocene age. The foraminiferal assemblage collected from the Stadium Conglomerate lies near the boundary between the middle and upper Eocene based upon the coccolith assemblage with which it is interbedded.

The Stadium Conglomerate is conformably overlain by richly fossiliferous strata of the Mission Valley Formation. The fossils include mollusks, calcareous nannoplankton, and mammals. The mollusks were collected from localities 28 through 33 (figure 7; plate 2A; and plate 2B, Kennedy and Peterson, 1975). These localities combined are indicated by M6 in figure 6. The fossil nannoplankton were collected from locality 15 (figure 8; plate 2B, Kennedy and Peterson, 1975). This locality is shown by N3 in figure 6. The fossil mammals have been collected from University of California, Riverside, locality UCR RV 7048, University of California, Berkeley, locality UCMP V6871 and Los Angeles County Museum locality LACM 65190 (figure 9; plate 3B, Kennedy and Peterson, 1975). Together these localities are indicated as V4 in figure 6.

The molluscan fauna collected from localities 28 through 32 include *Tellina tehachapii* Anderson and Hanna, *Macrocallista andersoni* Dickerson, and *Crassatella uvasana* s. s. Gabb. These species when considered together are characteristic of the upper Eocene Tejon Stage of California (Givens, 1974). Mollusks collected from locality 33 which is approximately 25 m stratigraphically higher in the section than locality 28 include *Turritella uvasana sargeanti* (Anderson and Hanna) which Givens (1973) considers restricted to the upper part of the Tejon Stage.

The calcareous nannoplankton flora collected from locality 15 is sparse but includes several distinctive species including *Reticulofenestra umbilica* (Levin) and *Discoaster distinctus* Martini, which together suggest either a late middle Eocene or early Eocene age.

The artiodactyl fauna collected from the Mission Valley Formation has been reported by Golz (1973) to belong to a stage of evolution correlative with the Uinta C.

The Mission Valley Formation is conformably overlain by rocks of the Pomerado Conglomerate and no fossils have been found in these rocks. A late Eocene age has been assigned to the Pomerado Conglomerate at its type locality on the basis of its superpositional and gradational relationship with the underlying fossiliferous Mission Valley Formation at its type locality.

The biostratigraphic relationship between fossil mollusks, calcareous nannoplankton, Foraminifera, and mammals with respect to the lithostratigraphy of the Eocene San Diego embayment is illustrated in figure 6. As discussed in the preceding pages, each of the fossil localities shown in figure 6 is a composite of many field localities that occur at or very near the same stratigraphic horizon. These composite localities have been plotted with respect to their relative vertical and horizontal stratigraphic position within the lithostratigraphic regime. The boundaries and postulated interrelationships of the West Coast (California) Marine Stages, European Stages, Epochs, Series, and absolute (radiometric) time scale in millions of years are also shown.

Composite mollusk locality M5 and composite mammal locality V2 lie within a few meters of the same stratigraphic level and are considered to be correlative in age. The molluscan fauna belongs to the "Transition Stage" and the mammalian fauna to the Uinta B Mammal Age of North America. Similarly, composite mollusk locality M6 and mammal locality V4 lie at the same stratigraphic interval. These are Tejon and Uinta C in age respectively. Givens (1974) has correlated the Tejon Stage of the southern California Ventura Basin with the upper Eocene of Europe, on the basis of species also reported here from composite locality M6. The calcareous nannoplankton from composite locality N3 which lies at the same stratigraphic interval as locality M6, are also suggestive of a late Eocene age.

Calcareous nannoplankton from composite fossil locality N1 within the middle part of the Ardath Shale have been correlated with the *Discoaster sublodoensis* zone (Bukry and Kennedy, 1969). The flora of this zone has previously been reported from the Canoas Siltstone Member of the Kreyenhagen Formation on Garza Creek near Oil City, California; from the middle Lutetian strata at Gibret, France; and from the Lutetian strata of the Paris basin in France (Bouche, 1962).

The fossil molluscan assemblage collected from composite locality M5 is indicative of the "Transition Stage". Givens (1974) has shown that the "Transition Stage" of southern California, as originally defined by Clark and Vokes (1936), overlaps the middle-upper Eocene boundary as established in the same strata by planktonic correlations with type Eocene strata in Europe.

Composite locality V2 also lies at the same stratigraphic interval at locality M5 and is therefore considered to likewise lie near the middle upper Eocene boundary. The fossils from composite localities F2 and N2 are stratigraphically higher than those from localities M5 and V2 and are considered to be from rocks that are at least late middle Eocene age. The fossils from composite localities M3, V1, and F1 lie at the same stratigraphic interval as those from N1 and are, therefore, also middle Eocene in age.

In conclusion, composite localities N3, M6, and V4 lie within the Mission Valley Formation and are late Eocene in age. Composite localities N1, M3, F1, and V1 lie within the Ardath Shale and are middle Eocene in age. The boundary between the middle and late Eocene lies intermediate between these two units within parts of the Scripps Formation, Friars Formation, and Stadium Conglomerate.

POST-EOCENE DEPOSITS

Miocene

Andesite Dike

An andesite dike is located approximately 600 m north of the Scripps Institution of Oceanography pier (plate 2A). The rock is black, fine grained, and has flow structures and columnar joints. The dike strikes approximately N. 45° E., but its intersection with Eocene rock in the sea cliff cannot be seen. The dike has been observed by scuba diving to extend from the beach directly beneath the U.S. Fishery Oceanography Center for a distance of approximately 400 m to the southwest (W. Reetz, personal communication, 1971). A whole-rock potassium-argon analysis of this rock, which shows some evidence of wall-rock assimilation, gave an age of 10.9 ± 1.1 million years (J.W. Hawkins, personal communication, 1970).

Pliocene and Pleistocene

Pliocene and Pleistocene rocks include marine sandstone and conglomerate of the upper Pliocene San Diego Formation, marine and nonmarine sandstone of the lower Pleistocene Lindavista Formation, and lagoonal and nonmarine sandstone of the upper Pleistocene Bay Point Formation.

San Diego Formation

The San Diego Formation (Dall, 1898) is middle or late Pliocene in age (Hertlein and Grant, 1944; Cleveland, 1960). It crops out from the lower south-facing slopes of Mount Soledad at Pacific Beach south to near San Diego Civic Center and along the north-facing slopes of Mission Valley from near Old Town to the eastern boundary of the area. These exposures, attaining a maximum thickness of 30 m, are composed of yellowish-brown, fine- to medium-grained, poorly indurated sandstone. Cobble conglomerate, thin beds of bentonite, marl, and brown mudstone further characterize the section. The thickness of the San Diego Formation increases markedly to the south, where it has been reported to attain a maximum thickness of 400 m (Hertlein and Grant, 1939). The lower 200 m of this section correlates to the south with the Miocene-Pliocene Rosarito Beach Formation in northern Baja California. The cobble conglomerate beds are composed primarily of Poway-type clasts, but some beds contain up to 50 percent clasts of granitic and metavolcanic rocks derived from the local basement complex. The bentonite is light brown, waxy to earthy, expansible, and very soft.

The San Diego Formation rests unconformably on the older pre-Pliocene rocks and is overlain by the Lindavista Formation. It is separated from the overlying Lindavista Formation in some places by an unconformity, but in other places the contact is gradational.

Lindavista Formation

The Lindavista Formation was named by Hanna (1926) for exposures at the Lindavista railroad siding in the La Jolla quadrangle (Lat 32° 53' N.; Long 117° 11' W.). The formation consists of near-shore marine and nonmarine sediments deposited on a 10 kilometer-wide wave-cut platform (Lindavista Terrace of Hanna, 1926) following the deposition of the middle or late Pliocene San Diego Formation (Hertlein and Grant, 1944) and prior to the deposition of the fossiliferous late Pleistocene (Sangamon) Bay Point Formation (Kern, 1971). A molluscan fauna from the Lindavista Formation, including the extinct species *Pecten bellus*, not known from the late Pleistocene, suggests an early Pleistocene or late Pliocene age for these rocks (G. Kennedy, 1973). The Lindavista Formation is predominantly composed of moderate reddish-brown interbedded sandstone and conglomerate. Ferruginous cement, mainly hematite, gives the Lindavista Formation its characteristic color and a resistant nature.

Both the coarse-grained and fine-grained rocks of the Lindavista Formation have been largely derived from the older sedimentary rocks within the San Diego embayment. Where iron staining, so common to the Lindavista Formation, extends downward into the underlying Eocene rocks, the two become difficult to differentiate.

Bay Point Formation

The Bay Point Formation (Hertlein and Grant, 1939) is widespread and well exposed in the area adjacent to the present-day coastline. It is composed mostly of marine and nonmarine, poorly consolidated, fine- and medium-grained, pale brown, fossiliferous sandstone.

The fossils found occur between 0 and 30 m above mean high tide and include mollusks, Foraminifera, and ostracods. These together indicate a brackish water estuarine depositional environment and a late Pleistocene (Sangamon) age (Kern, 1971).

The marine part of the Bay Point Formation interfingers with unfossiliferous sandstone that lies generally more than 30 m but less than 60 m above sea level. This part of the Bay Point Formation is considered to be nonmarine slope wash; however, it has not been differentiated on the geologic map.

Pleistocene and Holocene Surficial Deposits

The Pleistocene and Holocene surficial deposits are detrital materials which include stream-terrace, landslide, alluvium, slope wash, and beach deposits and artificially compacted fill.

Stream-Terrace Deposits

Stream-terrace deposits occur very locally as thin veneer along the larger drainage courses. The deposits include unconsolidated sand and gravel derived from older sedimentary, igneous, and metamorphic rocks.

Landslide Deposits

The study area is underlain in large part by incompetent sedimentary rocks which have been broadly dissected by shallow westward-flowing stream channels. Most of the landslides in the map area are rotational slumps and have occurred along valley walls where rocks of the Delmar, Friars, and Mission Valley Formations crop out.

Slope stability with respect to potential landsliding is dependent on several factors: (1) the strength of the rock material, (2) the slope angle, (3) the degree to which planar surfaces, such as bedding, joint, and fault planes, are dipping out of the slope, (4) the susceptibility of the slope-forming materials to saturation by water, which is related to the water source, permeability, porosity, and conditions of drainage.

The landslides are gravity slides resulting from basal erosion of oversteepened slopes, ground water saturation, surface-water erosion, and poorly consolidated rock. Sliding has generally occurred along a multiple slip surface associated with expansible clay The slides have consistently maintained internal homogeneity, and rotation of the slide mass is normally less than five degrees. Subsurface examination of these slides was beyond the scope of this study.

Most of the stream channels that dissect the soft sedimentary cover are strongly asymmetrical with their steep side exposed to the north. The north-facing slopes commonly stand 10 to 15 degrees steeper than the south-facing ones, which seldom reach angles greater than 30 degrees. Oversteepening of the stream channels is controlled in part by the presence of resistant impermeable rock layers exposed along the upper slopes as erosional ledges and platforms. The ledges protect the softer incompetent material below from direct rainfall but not from stream erosion. Landslides occur beneath the resistant conglomerate within incompetent rock as a result of undercutting by adjacent streams.

Several man-induced slides investigated during this study were found to occur beneath resistant conglomerate layers and within soft sandstone and claystone of the Delmar, Friars, and Mission Valley Formations just as in most of the natural slides. Stability filling (compacted fill placed over a benched cut slope) may be one means by which this type of failure can be avoided. Landslide incidence in the area is greatly increased during periods of high annual rainfall; good subdrainage may be another means of control. During periods of high precipitation, the saturation of bedrock may result in the lowering of the internal strength of perhaps already weak rock. Removal of slope-supporting material might then increase landslide potential.

The landslide deposits of the area can be subdivided into five major groups based on lithology and genesis. These are (1) rotational slump deposits associated with the shallow stream channels underlain by sedimentary rocks of Eocene age (photo 3); (2) rotational slump deposits associated with the sea cliffs underlain by sedimentary rocks of Eocene age north of the Rose Canyon fault zone (photo 4); (3) rotational slump deposits associated with sedimentary rocks of Upper Cretaceous and Eocene age within the Rose Canyon fault zone (photo 5); (4) rotational slump deposits associated with sedimentary rocks of Upper Cretaceous age on the east-facing slopes of the Point Loma Peninsula (photos 6

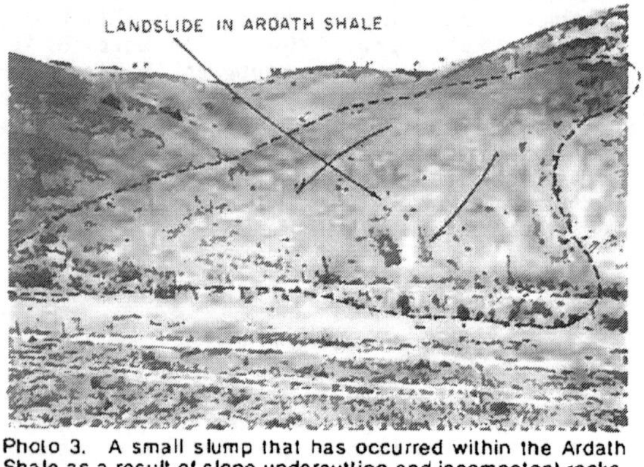

Photo 3. A small slump that has occurred within the Ardath Shale as a result of slope undercutting and incompetent rocks, looking southeast.

Photo 4. Torrey Pines State Park landslide, looking east. The landslide is located 2400 meters south of Soledad Valley in the sea cliffs. The slide mass is composed of incompetent rocks that are of Eocene age and belong to the Delmar Formation, Torrey Sandstone, and Ardath Shale.

and 7); (5) rockfall deposits associated with sedimentary rocks of Upper Cretaceous and Pleistocene age in the sea cliffs between the Rose Canyon fault zone and the tip of Point Loma Peninsula (photo 8; Kennedy, 1973b).

Eighteen clay samples were collected from formations that have high landslide incidence. These samples were analyzed for their particle size distribution, clay mineral content, and Atterberg limits.

Table 1 presents the results of the Atterberg tests and the quantity of the individual size fractions from the particle size distribution tests.

For engineering purposes the clay fraction is defined as less than 0.002 millimeter (mm), the silt fraction from 0.002 to 0.074 mm, and the sand fraction from 0.074 to 2.0 millimeters. The liquid limit (LL) is defined as the minimum moisture content at which the material behaves as a liquid using this test. The plastic limit (PL) is defined as the minimum moisture content at which the sample behaves plastically using this specified test procedure. The plasticity index (PI) is the numerical difference between the liquid limit and the plastic limit. The plasticity index is then the range of moisture content over which the sample behaves plastically. There exists a direct relationship between the liquid limit and "compression index," and between the liquid limit and the "coefficient of consolidation" (Terzaghi and Peck, 1968), whereas an inverse relationship exists between the plasticity index and "shear resistance."

Photo 5. Landslides that have occurred as a result of oversteepened slopes associated with an erosional scarp, looking southeast along the Mount Soledad fault.

The stated median particle size was taken from the 50 percent cumulative level of particle size distribution graphs. The arithmetic mean of the median particle sizes among all the samples is 0.016 mm (16 microns), which is in the lower portion of the silt range.

Associations between plasticity index, particle size, and the percentage of clay-sized particles are readily observed from the data of table 1. The medium-plasticity samples (PI = 10 to 20) have an average median particle size of 47 microns. High-plasticity samples (PI = 20 to 40) have an average median particle size of 14 microns. The very-high-plasticity samples (PI > 40) have a mean particle size of 6 microns.

Most landslides have occurred in rocks that have a plasticity index greater than 20. Surficial slumping of slopes underlain by rocks of the Delmar and Friars Formations and Ardath Shale is most abundant where the plasticity index is greater than 35. Expansion cracks are especially well developed and up to 50 centimeters (cm) deep and 5-10 cm wide in much of the area underlain by the Delmar and Friars Formations where the plastic index is greater than 40.

Photo 6. Ancient landslide deposits underlain by rocks of the Upper Cretaceous Rosario Group on Point Loma Peninsula, looking west.

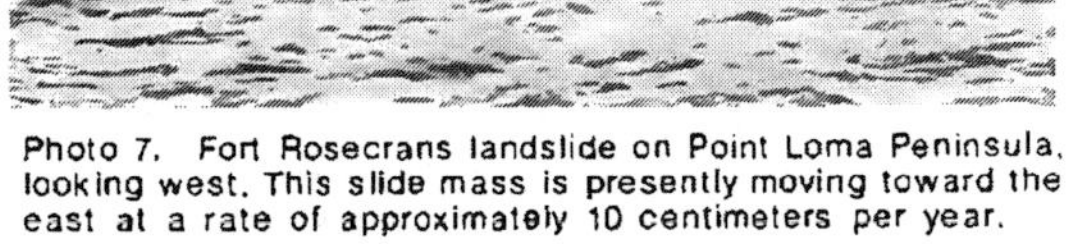

Photo 7. Fort Rosecrans landslide on Point Loma Peninsula, looking west. This slide mass is presently moving toward the east at a rate of approximately 10 centimeters per year.

Table 1. Clay mineral analyses.

Locality number	Locality description and remarks	Atterberg Limits			Particle Size Distribution				Mineralogy*					
		Liquid limit	Plastic limit	Plasticity index	% Sand	% Silt	% Clay	Median particle size in mm	Smectite	Mica	Kaolin	Quartz	Feldspar	Calcite
C-1	200 m north of Banana Street on Villa De La Valle Road in south facing cut 2 m above road level; lat. 32° 59′ 8″ N., long. 117° 14′ 43″ W.; Delmar Formation; 12.4Å with high background 20–40Å; rather low, broad reflections.	42	17	25	17	68	15	0.023	M**	Tr	Tr–Mr	Tr–Mr	Mr–Tr	Tr
C-2	670 m east of locality C-1 in road cut for Villa De La Valle Road, 2 m above road level; lat. 32° 59′ 10″ N., long. 117° 14′ 18″ W.; Delmar Formation; 12.6Å with very weak 25–26Å superlattice line.	57	23	34	13	56	31	0.0055	M	Tr	Tr–Mr	Mr	Tr	ND
C-3	150 m east of Puerta Del Sol Road in road cut on north side of Villa De La Valle Road 2 m above road level; lat. 32° 59′ 42″ N., long. 117° 12′ 38″ W., Delmar Formation; 17.6Å with poorly defined 26–28Å superlattice line.	41	16	25	10	72	18	0.027	M	ND	Tr	Mr–Tr	Tr	ND
C-4	300 m south of La Zanja Canyon on west side of road leading to Black Mountain Road in exploration trench 2 m above floor; lat. 32° 59′ 46″ N., long. 117° 9′ 12″ W.; Delmar and Friars Formations undifferentiated; extremely low, broad 13.4Å line; mostly x-ray amorphous (allophane); minor gypsum present.	71	22	49	23	47	30		Mr	Tr	Tr	Mr	Tr	Tr
C-5	10 m vertically above locality C-4; lat. 32° 59′ 46″ N., long. 117° 9′ 12″ W.; Delmar and Friars Formations undifferentiated; very broad, low intensity 12.8Å with high background; much noncrystalline (allophane) material present.	63	20	43	15	51	34		M	ND	Tr–Mr	Mr–M	Tr–Mr	ND
C-6	1,500 m north of Penasquitos River Valley drainage on west side of Black Mountain Road; lat. 32° 57′ 15″ N., long. 117° 17′ 33″ W.; Mission Valley Formation; very sharp 15Å with easily observable 30Å line.	55	19	36	10	67	23	0.0095	M	ND	Tr	Tr–Mr	Mr	Tr
C-7	1,000 m north of Penasquitos River drainage and 550 m south of locality C-6 in road cut on west side of Black Mountain Road, 1 m above road level; lat. 32° 56′ 58″ N., long. 117° 07′ 42″ W.; Delmar and Friars Formations undifferentiated; fairly sharp 15Å with well defined 28.5Å superlattice line.	74	24	50	7	58	35	0.0053	M	ND	Tr–Mr	Mr–Tr	Tr	ND

continued on page following....

Table 1. Clay mineral analyses (continued).

Locality number	Locality description and remarks	Atterberg Limits			Particle Size Distribution				Mineralogy*					
		Liquid limit	Plastic limit	Plasticity index	% Sand	% Silt	% Clay	Median particle size in mm	Smectite	Mica	Kaolin	Quartz	Feldspar	Calcite
C-8	300 m south of locality C-7 in road cut on west side of Black Mountain Road, 1 m above road level; lat. 32° 56′ 50″ N., long. 117° 07′ 43″ W.; Delmar and Friars Formations undifferentiated; fairly broad 15Å with very diffused 25–35Å line.	46	31	15	?	73?	19	0.013	M	ND	Tr–Mr	Mr	Tr–Mr	ND
C-9	215 m south of locality C-8 in road cut on west side of Black Mountain Road at road level; lat. 32° 56′ 42″ N., long. 117° 07′ 42″ W.; Delmar and Friars Formations undifferentiated; low intensity, broad 15°Å plus high background 20–40Å.	63	19	44	16	53	31	0.0053	M	Tr	Tr–Mr	M–Mr	Tr–Mr	ND
C-10	150 m east of intersection between Coast Highway 101 and Carmel Valley Road on north side of road cut, 2 m above road level; lat. 32° 56′ 23″ N., long. 117° 15′ 32″ W.; Delmar Formation; extremely low, broad 12.5Å line; mostly x-ray amorphous (allophane).	59	19	40	4	57	39	0.0037	Mr	Tr	Tr–Mr	Mr–Tr	Tr–Mr	ND
C-11	120 m east of sea-cliff-stairway in northern part of Torrey Pines State Park on south side of road leading from stairway to park headquarters, 2 m above road level in high north facing cut; lat. 32° 55′ 34″ N., long. 117° 15′ 25″ W.; Delmar Formation; no line detected; x-ray amorphous (allophane).	56	22	34	0	54	46	0.0023	M	Mr–Tr	Mr–Tr	Tr	ND	ND
C-12	900 m south of Carmel Valley Road on Sorrento Valley Road in west facing road cut, 2 m above road level; lat. 32° 55′ 16″ N., long. 117° 14′ 17″ W.; Delmar Formation; essentially x-ray amorphous (allophane).	46	31	15	8	71	21	0.0082	M	Mr	Tr	Mr–M	Mr–M	Tr
C-13	First road cut for Interstate Highway 5 north of Penasquitos Valley, 1 m above highway level and 15 m north of the southern end of the cut; lat. 32° 54′ 30″ N., long. 117° 13′ 28″ W.; Ardath Shale.	53	33	20	26	74	0	0.006	Mr–Tr	Mr–Tr	Tr	Mr	Mr	ND
C-14	35 m north of intersection between Genesee Avenue and Interstate Highway 5 in road cut above onramp for freeway, 1 m above road level; lat. 32° 53′ 13″ N., long. 117° 13′ 33″ W.; Ardath Shale.	42	21	21	19	78	3	0.012	Mr–Tr	Mr	Tr	Mr	Mr	ND

continued on page following....

Table 1. Clay mineral analyses (continued).

Locality number	Locality description and remarks	Atterberg Limits			Particle Size Distribution				Mineralogy*					
		Liquid limit	Plastic limit	Plasticity index	% Sand	% Silt	% Clay	Median particle size in mm	Smectite	Mica	Kaolin	Quartz	Feldspar	Calcite
C-15	400 m north of Scripps Institution of Oceanography pier in sea cliff adjacent to small canyon, 3 m above mean high tide; lat. 32° 52′ 18″ N., long. 117° 15′ 6″ W.; Ardath Shale.	51	30	21	16	48	36	0.022	Mr–Tr	Mr	Mr–Tr	Mr	Mr	ND
C-16	Road cut at intersection of freeway offramp for Interstate Highway 5 and Gilman Drive, 3 m above road level, 60 m west of intersection; lat. 32° 51′ 02″ N., long. 117° 14′ 05″ W.; Ardath Shale.	43	23	20	24	65	11	0.010	M	Mr–Tr	Tr–Mr	Mr	Mr	ND
C-17	1,200 m west of U.S. Highway 395 on Friars Road, 10 m below contact between cobble conglomerate above (Stadium Conglomerate) and green claystone below (Friars Formation); lat. 32° 46′ 8″ N., long. 117° 10′ 13″ W.; Friars Formation; 12.6Å low, broad line with high 20–40Å background.	45	17	28	20	67	13	0.025	M	Tr	Mr	Tr	ND	ND
C-18	900 m west of U.S. Highway 395 on Friars Road, 8 m stratigraphically above locality C-17; lat. 32° 46′ 10″ N., long. 117° 10′ 13″ W.; Friars Formation; 12.4Å fairly sharp but low intensity line; no superlattice line or high background indicating mixed layer.	60	20	40	3	81	16	0.014	M	ND	Mr	Tr	Tr	ND

* Mineral estimates based on diffractogram line intensity. Analyses by Paul Anderson, California Division of Mines and Geology.
** M=major, over 25%; Mr=minor, 5–25%; Tr=trace, less than 5%; ND=not detected.

Photo 8. Sunset Cliffs located on the northern part of the Point Loma Peninsula, looking east. The arcuate coastline development here is the result of rockfall landsliding.

Alluvium and Slope Wash

Alluvium consists primarily of poorly consolidated stream deposits of silt, sand, and cobble-sized particles derived from bedrock sources that lie within or near the area. These deposits intertongue with Holocene slope wash that commonly mantles the lower valley slopes throughout coastal San Diego County. Alluvium and slope wash are mostly undifferentiated on the geologic maps.

Slope wash deposits are poorly consolidated surficial materials derived chiefly from nearby soil and decomposed bedrock sources. The slope wash is deposited along the flanks of the lower valley slopes by the actions of gravity and surface water. Thick deposits of slope wash are especially common on the Delmar Formation, Ardath Shale, Scripps Formation, and Friars Formation where deep soil horizons have developed. Expansive clay materials deposited as slope wash yield the hummocky topography developed on rocks of lagoonal and nonmarine origin.

Beach Deposits

The beach deposits are composed of unconsolidated sand and silt. They mantle those parts of the present day sea coast where erosional conditions are slow. They are derived from many sources as a result of longshore drift and alluvial discharge from the major stream courses.

Artificially Compacted Fill

Artificial fill consists of artificially compacted earth materials derived from many sources. Only large areas underlain by artificial fill have been delineated on the geologic maps.

STRUCTURE AND SEISMIC HISTORY

Faults in the San Diego coastal area lie within a regional northwest-striking right-lateral fault system that includes the active Mission Creek, San Andreas, San Jacinto, and Elsinore fault zones to the east and the Agua Blanca, San Clemente, and Rampart fault zones to the west. Within the study area the faults comprise two prominent sets. One set strikes parallel to the regional grain, whereas the other set strikes northeast. Faults belonging to both of these sets have displaced rocks of the late Pleistocene Bay Point Formation which has, on the basis of a rich nearshore molluscan fauna, been assigned an age of approximately 100,000 years (Kern, 1971).

The most prominent faults within the northwest striking set belong to the Rose Canyon fault zone.

These faults juxtapose nearly flat-lying Eocene, Pliocene, and Pleistocene rocks with steeply tilted Upper Cretaceous and Eocene rocks. The Rose Canyon fault has been considered a southern extension of the Newport-Inglewood fault zone (Corey, 1954; Emery, 1960; King, 1969; Moore, 1972) and a northern extension of both the Los Buenos and the San Miguel faults (Wiegand, 1970; Moore and Kennedy, 1970; Moore, 1972). If the Rose Canyon-Los Buenos fault zone is continuous, it has an onshore length of 65 km and extends from La Jolla on the north to near El Rosarita in northern Baja California on the south. If the Rose Canyon fault zone is continuous with the San Miguel fault, the onshore length of this segment is over 250 km and extends from La Jolla on the north to near San Miguel in northern Baja California on the south. The mapped northern offshore extension of the Rose Canyon fault zone extends from La Jolla to within 45 km of the southern onshore termination of the Newport-Inglewood fault zone (Moore, 1972). Point Loma and Mount Soledad are fault blocks that were uplifted along the Rose Canyon fault zone in part after the deposition of the Lindavista Formation. At least 135 m of vertical separation can be measured in the vicinity of Mount Soledad and La Jolla (plate 2A). The direction of the vertical movement at La Jolla is west side up, whereas at Mission Bay it is west side down (sections B-B', C-C'). At least 100 m of separation has been shown on the Rose Canyon fault zone in the Mission Bay area with the west side down (Peterson, 1970). Possibly the Mount Soledad block rotated along an axis normal to the strike of the fault zone thereby elevating Mount Soledad along the northwest side of the zone and sinking Mission Bay along the southwest side. This tilted block model is supported by the fact that a 30 m high wave-cut platform upon which the Lindavista Formation originally was deposited, on the south-facing slopes of Mount Soledad, is inclined to the south and extends nearly continuously from an elevation of over 250 m at the north to near sea level at the south.

Steep folds associated with the northeasternmost part of the Mount Soledad block (section B-B') are considered to be in part the result of compression developed by strike-slip movement. The rotation of Mount Soledad was questionably caused by flexure associated with the change in strike of the Rose Canyon and Mount Soledad faults between La Jolla Cove and Rose Canyon (plate 2A). The Tertiary and younger rocks that lie immediately west of the fault zone are not deformed and they make a strike-slip model and compressional folding logical. It has been suggested that the Rose Canyon fault is part of a regional right-lateral strike-slip fault system. The distribution of the San Diego Formation along the Rose Canyon fault zone between Pacific Beach and Tecolote Canyon(plate 2A) is interpreted as resulting from 4 km of right-lateral strike-slip motion on the Rose Canyon fault. Horizontal slickensides measured on the Rose Canyon fault at La Jolla and Mount Soledad further suggest strike-slip faulting.

In the northern coastal part of the area at Torrey Pines State Park and in the southern part of the area at Point Loma, the structural grain of the area is nearly perpendicular to the Rose Canyon fault zone. The average strike of the grain is 30 to 40 degrees east of north. The separation along most of the faults is vertical and ranges from 1 cm to 100 m.

Several northeast striking faults displace Pleistocene and younger(?) deposits. The Carmel Valley fault south of Del Mar (plate 1A) has approximately 2 m of vertical separation, involving the Lindavista Formation at lat 32° 55' 10" N.; long 117° 14' 50" W. Along this same fault at lat 32° 55' 20" N.; long 117° 14' 50" W., the late Pleistocene Bay Point Formation has been tilted approximately 10 degrees. It is speculated, however, that at least part of this dip is initial and related to its deposition. A small fault on the east side of the Point Loma Peninsula, located at lat 32° 40' 40" N.; long 117° 14' 15" W., also displaces rocks of the Bay Point Formation. The separation on this fault is dip-slip and on the order of 3 meters. The Bay Point Formation is also faulted by a small northeast striking fault that intersects the Point Loma fault at an oblique angle(plate3A). The vertical separation of the Bay Point Formation related to this fault and the Point Loma fault together is in excess of 30 meters.

The possibility of Holocene fault activity in the area is not ruled out, though no direct field evidence supports this fact. Holocene faulting is indirectly postulated by the fact that historic seismicity might be related to faulting on the Rose Canyon fault zone in the San Diego Bay area and by subbottom acoustic profiles that show probable Holocene sediments offset on the sea floor at a point approximately 25 km north of La Jolla along the trace of the Rose Canyon fault (Moore, 1972).

Forty-four earthquakes of Richter magnitude 2.5 to 3.7 (M 2.5-3.7) have been recorded within the greater San Diego area by the California Institute of Technology Seismological Laboratory since 1950. Three of these which occurred in the vicinity of San Diego Bay on June 21 and 22 and July 14, 1964, had epicenter localities within a few kilometers of San Diego and magnitudes of 3.7, 3.6, and 3.5, respectively.

In addition to earthquakes originating in the San Diego area, ground shaking has been felt there initiated by earthquakes that have had epicenters up to 100 km away. Several of these earthquakes have caused damage in San Diego and are worthy of mention.

The 1933 Long Beach earthquake (M 6.6) caused minor damage throughout northwestern San Diego County and was felt sharply as far south as San Diego. The epicenter is shown by the California Institute of Technology Seismological Laboratory to have been south of Long Beach along the inferred trace of the Newport-Inglewood fault zone.

On November 4, 1949, and on February 9, 1956, earthquakes felt sharply in San Diego occurred on the Valecitos-San Miguel fault in northern Baja California. The 1949 earthquake (M 5.7) had an epicentral distance of approximately 75 km southeast of the San Diego Civic Center. The epicenter of the 1956 earthquake (M 6.8) was approximately 175 km south of San Diego. Three aftershocks of this earthquake, with magnitudes greater than 6, occurred in 1956 on February 9, 14, and 15. Ground rupture of 20 km was associated with this earthquake in the vicinity of the epicenter (Shor and Roberts, 1958).

Several earthquakes (M > 5) have been recorded on the Agua Blanca fault in northern Baja California during the past 30 years with epicentral distances within 125 km of San Diego. Holocene fault scarps in the area between Ensenada and Santo Tomas suggest surface rupture has occurred in at least the past few thousand years.

During the past 35 years, earthquakes located in the Imperial Valley and Salton Trough area were felt in western San Diego County. Three of these—the 1940 Imperial Valley earthquake (M 7.1), the 1951 Superstition Hills earthquake (M 5.6), and the 1968 Borrego Mountain earthquake (M 6.5)—caused minor damage to structures in the San Diego coastal area and initiated landsliding of sea cliff property at Point Loma, La Jolla, and Torrey Pines State Park (Kennedy, 1973).

The February 9, 1971, San Fernando earthquake (M 6.4) was felt sharply throughout the southern California coastal area. The intensity of the earthquake at San Diego was V (Scott, 1971) on the Modified Mercalli scale. Minor damage was reported as far south as the Mexican border, and two small landslides occurred at Sunset Cliffs as a direct result of the initial shock.

Allen *et al.* (1965) show a strain release of 0.25 to 4 (M 3) earthquakes/100 km^2 for the 29-year period between 1934 and 1963 for the San Diego area.

REFERENCES CITED

Allen, C.R., *et al.*, 1965, Relationship between seismicity and geologic structure in the southern California region: Seismological Soc. of America Bull., v. 55, no. 4, p. 753-797.

Allison, E.C., 1964, Geology of areas bordering Gulf of California: American Assoc. Petroleum Geologists Mem. 3, p. 3-29.

Beal, C.H., 1924, Informe sobre la exploracion geologica de la Baja California: Boletin Petroleo, v. 17, p. 417-473.

Berggren, W.A., 1969, Rates of evolution in some Cenozoic planktonic Foraminifera: Micropaleo, v. 15, no. 3.

Bouche, P.M., 1962, Nannofossiles calcaires du Lutetien du bassin de Paris: Rev. Micropaleontologie, v. 5, p. 75-103.

Bukry, David, and Kennedy, M.P., 1969, Cretaceous and Eocene coccoliths at San Diego, California, *in* Short contributions to California geology: California Division of Mines and Geology Special Report 100, p. 33-43.

Bushee, J., Holden, J., Geyer, B., and Gastil, G., 1963, Lead-alpha dates for some basement rocks of southwestern California: Geol. Soc. America Bull., v. 74, p. 803-806.

Clark, B.L., and Vokes, H.E., 1936, Summary of marine Eocene sequence of western North America: Geol. Soc. America Bull., v. 47, p. 851-876, 2 pls., 3 figs.

Cleveland, G.B., 1960, Geology of the Otay clay deposit, San Diego County, California: California Div. Mines Spec. Rept. 64, p. 16.

Corey, W.H., 1954, Tertiary basins of southern California, *in* Geol. of southern California: California Div. Mines Bull. 170, chpt. 3, p. 73-83.

Dall, W.H., 1898, 18th Ann. Rept.: U.S. Geol. Survey, pt. 2, correlation table opp. p. 334.

DeLisle, M., Morgan, J. R., Heidenbrand, J., and Gastil, G., 1965, Lead-alpha ages and possible sources of metavolcanic rock clasts in the Poway Conglomerate, southwest California: Geol. Soc. America Bull., v. 76, p. 1069-1074.

Ellis, A.J., 1919, Geology, western part of San Diego County, California: U.S. Geol. Survey Water-Supply Paper 446, p. 50-76.

Emery, K.O., 1960, The sea off southern California, a modern habitat of petroleum: New York, London, John Wiley, 366 p., map.

Evernden, J.F., and Kistler, R.W., 1970, Chronology of emplacement of Mesozoic batholithic complexes in California and western Nevada: U.S. Geol. Survey Prof. Paper 623, 42 p.

Fife, D.L., Minch, J.A., and Crampton, P.J., 1967, Late Jurassic Age of the Santiago Peak Volcanics, California: Geol. Soc. America Bull., v. 78, p. 299-304.

Flynn, C.J., 1970, Post-batholithic geology of the La Gloria-Presa Rodriquez area, Baja California, Mexico: Geol. Soc. America Bull., v. 81, p. 1789-1806.

Givens, C.R., 1974, Eocene Molluscan biostratigraphy of the Pine Mountain area, Ventura County, California: University of California, Dept. Geol. Sci. Bull., v. 109, 107 p.

Golz, D.J., 1971, Selenodont Artiodactyls from the Eocene of southern California: Geol. Soc. America Abstracts with Program, v. 6, p. 125.

Golz, D.J., 1973, The Eocene Artiodactyla of southern California: Unpublished Ph.D. dissertation, University of California, Riverside.

Golz, D.J., and Kennedy, M.P., 1971, Comparison of Mammalian and Invertebrate Chronologies in the Eocene of southern California: Geol. Soc. America Abstracts with Program, v. 6, p. 125.

Gray, C.H., Jr., Kennedy, M.P., and Morton, P.K., 1971, Petroleum potential of southern coastal and mountain area, California: American Assoc. Petroleum Geologists, Mem. 15, p. 372-383.

Hanna, M.A., 1926, Geology of the La Jolla quadrangle, California: University of California, Dept. Geol. Sci. Bull., v. 16, p. 187-246.

Hertlein, L.G., and Grant, U.S., IV, 1939, Geology and oil possibilities of southwestern San Diego County: California Jour. Mines and Geol., v. 35, p. 57-78.

Hertlein, L.G., and Grant, U.S., IV, 1944, The geology and paleontology of the marine Pliocene of San Diego, California, pt. 1, Geology: San Diego Soc. Nat. History Mem., v. 2, p. 1-72.

Jenkins, D.G., 1965, Planktonic Foraminiferal Zones and New Taxa from the Danian to lower Miocene of New Zealand: New Zealand Jour. Geol. Geophys., v. 8, p. 1088-1126.

Kennedy, G.L., 1973, Early Pleistocene Invertebrate Faunule from the Lindavista Formation, San Diego, California: San Diego Soc. Nat. History Transactions, v. 17, p. 119-128.

Kennedy, M.P., 1967, Preliminary report, engineering geology of the city of San Diego, California: California Div. of Mines and Geology open file, 21 p., 3 maps, scale 1:24,000.

Kennedy, M.P., 1969, Preliminary geologic maps of portions of San Diego city, California: California Division of Mines and Geology open file reports 69-1 (Del Mar sheet), 68-10 (Del Mar - La Jolla sheet), 69-13 (La Jolla - Point Loma sheet), 69-14 (Point Loma sheet), scale 1:9,6000.

Kennedy, M.P., 1971, Eocene shoreline facies in the San Diego coastal area, California: Geol. Soc. America Abstracts with Program, v. 6 p. 142.

Kennedy, M.P., 1973a, Stratigraphy of the San Diego embayment, California: Unpublished Ph.D. dissertation, University of California, Riverside.

Kennedy, M.P., 1973b, Sea cliff erosion at Sunset Cliffs, San Diego, California: California Div. Mines and Geology, California Geology, v. 26, p. 27-31.

Kennedy, M.P., and Moore, G.W., 1971a, Stratigraphic relations of upper Cretaceous and Eocene formations, San Diego coastal area, California: American Assoc. Petroleum Geologists Bull., v. 55, p. 709-722.

Kennedy, M.P., and Moore, G.W., 1971b, Stratigraphy and structure of the area between Oceanside and San Diego, California: geologic road log, *in* Elders W.A., ed., 1971, Geological excursions in southern California: Geol. Soc. America Cordilleran Section Meeting, Riverside, California, field trip guidebook.

Kennedy, M.P., and Peterson, G.L., 1975, Geology of the La Mesa, Poway, and SW 1/4 Escondido quadrangles, eastern San Diego metropolitan area, California: California Div. Mines and Geology Bull. 200B.

Kern, J.P., 1971, Paleoenvironmental analysis of a late Pleistocene estuary in southern California: Journal Paleo., v. 45, p. 810-823.

King, P.B., 1969, The tectonics of North America: U.S. Geol. Survey Prof. Paper 628, 94 p., map scale 1:5,000,000.

Larsen, E.S., 1948, Batholith and associated rocks of Corona, Elsinore, and San Luis Rey quadrangles, southern California: Geol. Soc. America Mem. 29, 182 p.

Mallory, V.S., 1959, Lower Tertiary biostratigraphy of the California coast ranges: American Assoc. Petrol. Geol., 297 p.

Merriam, R., 1968, Geologic reconnaissance of northwest Sonora: Stanford University Pub. Geol. Sci., v. 11, p. 287.

Milow, E.D., and Ennis, D.B., 1961, Guide to geologic field trip of southwestern San Diego County: Geol. Soc. America, Cordilleran Sec., 57th Ann. Mtg., Guidebook, p. 23-43.

Minch, J.A., 1967, Stratigraphy and structure of the Tijuana-Rosarito Beach area, northwestern Baja California, Mexico: Geol. Soc. America Bull., v. 78, p. 1155-1178.

Minch, J.A., 1972, The late Mesozoic-early Tertiary framework of continental sedimentation, northern Peninsular Ranges, Baja California, Mexico: Unpublished Ph.D. dissertation, University of California, Riverside.

Moore, E.J., 1968, Fossil mollusks of San Diego County: San Diego Soc. Nat. History Occasional Paper 15, 76 p.

Moore, G. W., 1972, Offshore extension of the Rose Canyon fault, San Diego, California: U.S. Geol. Survey Prof. Paper 800-C. p. C113-C116.

Moore, G.W., and Kennedy, M.P., 1970, Coastal geology of the California—Baja California border area: American Assoc. Petroleum Geologists Guidebook, Pacific Sections, fall field trip p. 4-9.

Morton, P.K., 1972, Geologic guidebook to the northern Peninsular Ranges, Orange and Riverside Counties, California: Prepared jointly by National Assoc., of Geol. Teachers and South Coast Geological Soc., for N.A.G.T. far western section meeting, Chapman College, Orange, California.

Nordstrom, C.W., 1970, Lusardi Formation—a post-batholithic Cretaceous conglomerate north of San Diego, California: Geol. Soc. America Bull., v. 81, p. 601-605.

Peterson, G.L., 1970, Quaternary deformation of the San Diego area, southwestern California: American Assoc. Petroleum Geologists Guidebook, Pacific Section, fall field trip, p. 120-126.

Peterson, G.L., and Kennedy, M.P., 1974, Lithostratigraphic variations in the Poway Group near San Diego, California: San Diego Soc. Nat. History Transactions, v. 17, p. 251-258.

Peterson, G.L., and Nordstrom, C.E., 1970, Sub-La Jolla unconformity in the vicinity of San Diego, California: American Assoc. Petroleum Geologists Bull., v. 54, p. 256-274.

Popenoe, W.P., Imlay, R.W., and Murphy, M.A., 1960, Correlation of the Cretaceous formations of the Pacific coast (United States and northwestern Mexico): Geol. Soc. America Bull., v. 71, p. 1491-1540.

Scott, N.H., 1971, Preliminary report on felt area and intensity, *in* The San Fernando, California, earthquake of February 9, 1971: U.S. Geol. Survey Prof. Paper 733, p. 153-154.

Sears, J.D., Hunt, C.B., and Hendricks, T.A., 1941, Transgressive and regressive Cretaceous deposits in southern San Juan Basin, New Mexico: U.S. Geol. Survey Prof. Paper 193-F, p. 100-121.

Shor, G.G., and Roberts, E., 1958, San Miguel, Baja California Norte, earthquakes of February, 1956; a field report: Seismological Society of America Bull., v. 48, p. 101-116.

Slitter, W.V., 1968, Upper Cretaceous Foraminifera from southern California and northwestern Baja California, Mexico: The University of Kansas Pubs., Art. 7, Ser. no. 49, p. 141.

Steineck, P.L., and Gibson, J.M., 1971, Age and correlation of the Eocene Ulatisian and Narizian stages, California: Geol. Soc. America Bull., v. 82, p. 477-480.

Strand, R.G., 1961, Geologic map of California—San Diego-El Centro sheet: California Div. Mines and Geology.

Terzaghi, K., and Peck, R.B., 1967, Soil mechanics in engineering practice: Wiley and Son, New York, 729 p.

Turner, H.C., Ebert, E.E., and Given, R.R., 1968, The Marine environment offshore from Point Loma, San Diego County: California Department of Fish and Game, Fish Bull. 140.

Weber, F.H., Jr., 1963, Geology and mineral resources of San Diego County, California: California Div. Mines and Geology County Rept 3, 309 p.

Wiegand, J.W., 1970, Evidence of a San Diego Bay-Tijuana fault: Assoc. of Engineering Geologists Bull., v. 7, p. 107-121.

Wood, H.E., Chaney, R.W., Clark, J., Colbert, E.H., Jepsen, G.L., Reeside, J.B., Jr., and Stock, C., 1941, Nomenclature and correlation of the North American Continental Tertiary: Geol. Soc. America Bull., v. 52, p. 1-48.

Woodford, A.O., Welday, E.E., and Merriam, R., 1968, Siliceous tuff clasts in the upper Paleogene of southern California: Geol. Soc. America Bull., v. 79, p. 1461-1486.

SECTION B

La Mesa, Poway,
and SW1/4 Escondido
quadrangles

ABSTRACT

The La Mesa, Poway, and SW1/4 Escondido 7.5-minute quadrangles cover approximately 380 square kilometers (km^2) within the eastern and northeastern San Diego metropolitan area. The geology of the area consists of two principal rock units: 1) an igneous and metamorphic basement complex and 2) a superjacent sedimentary succession of strata.

The basement complex is composed of the Upper Jurassic Santiago Peak Volcanics, a structurally complex, mildly metamorphosed unit composed of andesitic volcanic and volcaniclastic rocks and mid-Cretaceous rocks of the southern California batholith. Metamorphism of the Santiago Peak Volcanics, emplacement of the batholithic rocks, uplift, and carving of an erosion surface with relief in excess of 500 meters (m) was completed by Late Cretaceous time.

The post-batholithic superjacent sedimentary succession was deposited on this high relief erosion surface. Mapped stratal units include the Upper Cretaceous Lusardi Formation; the Eocene Friars Formation, Stadium Conglomerate, Mission Valley Formation, and Pomerado Conglomerate; the Pliocene San Diego Formation; the Pleistocene Lindavista Formation; and Holocene landslide, alluvial, slope wash and stream-terrace deposits.

The most abundant rocks of the sedimentary succession are middle and upper Eocene fossiliferous strata of marine, lagoonal, and nonmarine origin related to two major transgressive and regressive depositional episodes. The Lindavista Formation caps the older rocks throughout much of the area and was deposited during a marine regressional event following a marine planation. Later uplift of these deposits is indicated by the presence of fossiliferous marine strata that now lie at an elevation of 165 m. above sea level.

Seismically the area is quiet; however, it lies within a part of southern California considered to be a region of tectonic activity. Forty-four earthquakes of Richter magnitude between 2.5 and 3.7 (M 2.5 to M 3.7) have been recorded by the California Institute of Technology Seismological Laboratory since 1950 that have had epicentral localities within the greater San Diego metropolitan area.

No known Holocene faults exist in the area. However, the Pleistocene Lindavista Formation has been faulted in several places. One of these faults lies within the Mission Valley River drainage, south of Mission Gorge, where the Lindavista Formation has been offset at least 20 meters. A fault near Collwood Boulevard and Montezuma Road in the southern part of the area also cuts rocks of the Lindavista Formation and is considered to be a northern branch of the La Nacion fault.

Sand and gravel deposits useable for concrete, bituminous, and ceramic aggregate underlie a large part of the area. Clay deposits useable for ceramics, fire clay, and possibly lightweight aggregate are also abundant but have not been exploited. These widespread clay deposits are expansible, and closely associated with landslides, especially in those areas directly underlain by the Friars Formation.

The landslides mapped are rotational slumps that have occurred as the result of incompetent rock, saturation, expansible clay, and oversteepened valley slopes. Surficial landslides are associated with slopes steeper than 25 degrees that are underlain by rocks of the Friars and Mission Valley Formations throughout the area.

GEOLOGY OF THE EASTERN SAN DIEGO METROPOLITAN AREA, CALIFORNIA

La Mesa, Poway, and SW1/4 Escondido quadrangles

by Michael P. Kennedy[1] and Gary L. Peterson[2]

INTRODUCTION

In 1965 the California Division of Mines and Geology, in cooperation with the City of San Diego, began a comprehensive geologic investigation of the greater San Diego metropolitan area. This report deals with the geology of the eastern half of that investigation. A similar report has been written on the western half and includes the geology of the Del Mar, La Jolla, and Point Loma quadrangles (Kennedy, 1975).

The La Mesa, Poway, and SW 1/4 Escondido quadrangles comprise more than 25 percent of the greater San Diego metropolitan area (figure 1). This area is underlain by San Diego's richest sand, gravel, and crushed stone resources, deemed feasibly extractable in today's market for use in the Mission Valley, Mira Mesa, Poway, and Escondido suburbs. These resources and others, including rich clay deposits, are being rapidly covered by urban development. The clay deposits, which are locally expansible, in turn constitute a serious geologic hazard to development.

The geologic mapping and detailed descriptions of the rock units are intended to be used as aids in planning for land use and future development. The stratigraphic relationship between the rock units underlying the study area and those that underlie the area to the west (discussed by Kennedy, 1975) are shown in figure 2.

Previous geologic investigations that have been especially useful in this study include a ground water investigation by A.J. Ellis (1919), a stratigraphic and paleontologic study of the La Jolla quadrangle by M.A. Hanna (1926), two papers on geology and paleontology of the San Diego area by L.G. Hertlein and U.S. Grant IV (1939, 1944), and a monograph on the mineral resources of San Diego County by F.H. Weber, Jr. (1963).

The authors would like to extend special thanks to D.M. Morton and G.W. Moore of the United States Geological Survey for their suggestions and contributions pertinent to the results of this study. Acknowledgment is due also to M.O. Woodburne, P.K. Morton, G.B. Cleveland, F.H. Weber, Jr., C.H. Gray, Jr., M.A. Murphy, J.P. Kern, R.G. Strand, and Y.H. Smitter for their enthusiastic help, interesting discussions, and review of the maps and manuscript.

PRE-EOCENE DEPOSITS

Basement Complex

The basement complex consists of two principal rock units: (1) the Upper Jurassic Santiago Peak Volcanics, a succession of deformed and metamorphosed volcanic, volcaniclastic, and sedimentary rocks; and (2) mid-Cretaceous plutonic rocks of the southern California batholith, which intrude the Santiago Peak Volcanics.

[1]Geologist, California Division of Mines and Geology
[2]San Diego State University

Santiago Peak Volcanics

The Santiago Peak Volcanics comprise an elongate belt of mildly metamorphosed volcanic, volcaniclastic, and sedimentary rocks that crop out from the southern edge of the Los Angeles basin southward into Mexico (Gray *et al.*, 1971). These rocks were mapped in the San Diego area by Hanna (1926, p. 199-204) as "Black Mountain Volcanics," but that name was pre-empted, and Larsen (1948) suggested the substitute--Santiago Peak Volcanics.

The volcanic rocks range in composition from basalt to rhyolite but are predominantly dacite and andesite. The succession is typified by a wide variety of breccia, agglomerate, volcanic conglomerate, and fine-grained tuff-breccia. Highly silicified rock (probably tuff) and a variety of dark, dense, fine-grained hornfels occur locally. To the west, some local, thin, fossil-bearing marine sedimentary rocks are interbedded with the volcanic and volcaniclastic rocks. Included with the Santiago Peak Volcanics are a number of small plutons of mildly metamorphosed gabbro. These are herein included with the Santiago Peak Volcanics because they are metamorphosed and were probably feeders for the volcanic rocks rather than parts of the batholith.

The Santiago Peak Volcanics are hard and extremely resistant to erosion and form topographic highs. Most of the volcanic rocks are dark greenish gray where fresh but weather grayish red to dark reddish brown. The soil developed on the Santiago Peak Volcanics is the color of the weathered rock and supports the growth of dense chaparral.

Within a narrow 2-kilometer-long belt, approximately 1 km northwest of Rancho Bernardo in the Escondido quadrangle (plate 1B), a succession of low-grade metamorphic slate and quartzite crops out. These rocks were considered by Larsen (1948) to belong to the Bedford Canyon Formation. Within the northeastern part of the Poway quadrangle, a similar succession was included by Hanna within his "Black Mountain Volcanics." In this study the rocks at both of these exposures have been included in the Santiago Peak Volcanics. Although they differ somewhat from more characteristic Santiago Peak Volcanics, the difference is not deemed enough to correlate them with the Bedford Canyon Formation.

Age estimates for the Santiago Peak Volcanics have ranged from Late Triassic (Hanna, 1926) to mid-Cretaceous (Milow and Ennis, 1961). However Fife *et al.* (1967) showed them to be latest Jurassic (Portlandian) based on fossils from sedimentary interbeds.

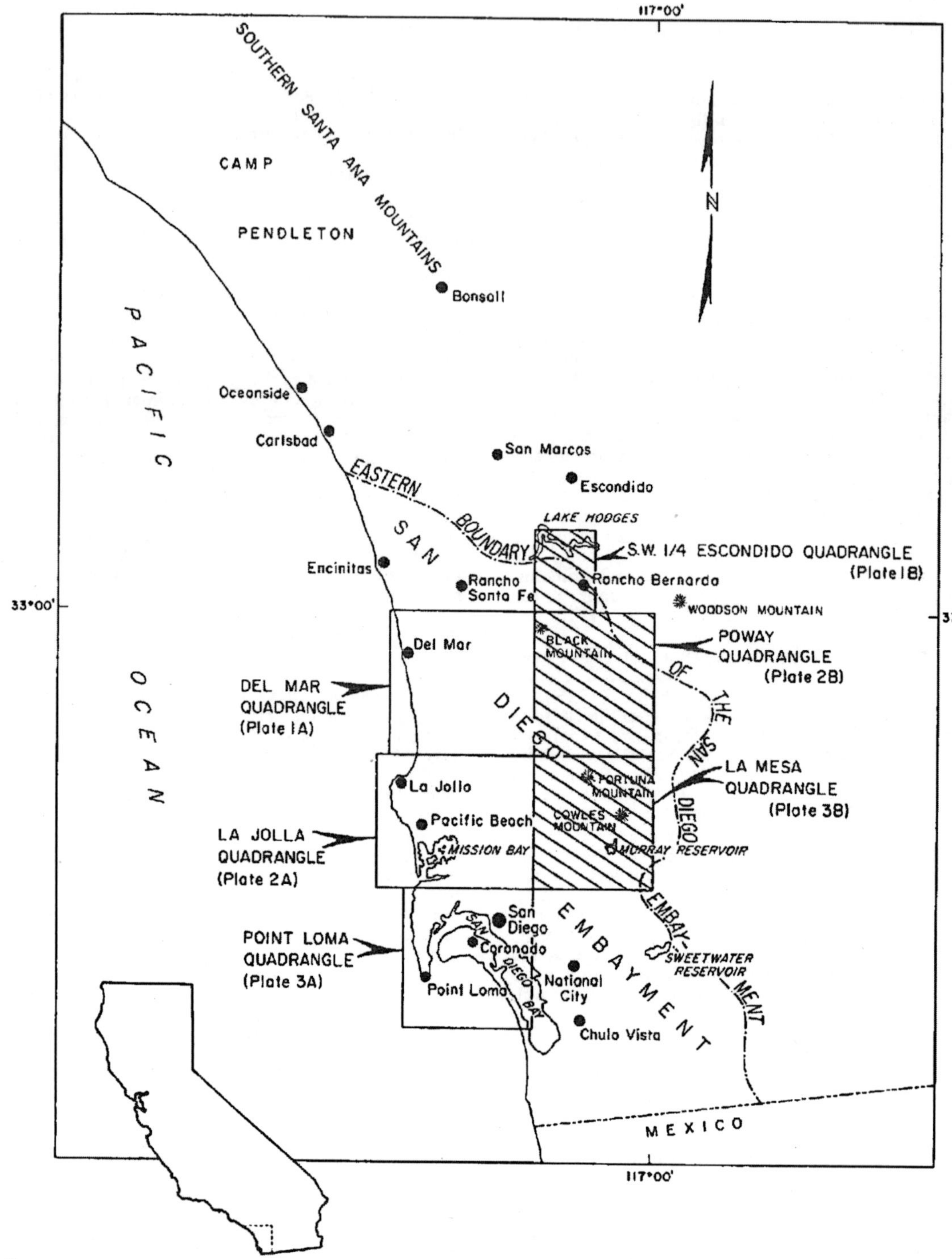

Figure 1. Index map showing the location of the La Mesa, Poway, and SW 1/4 Escondido 7.5-minute quadrangles.

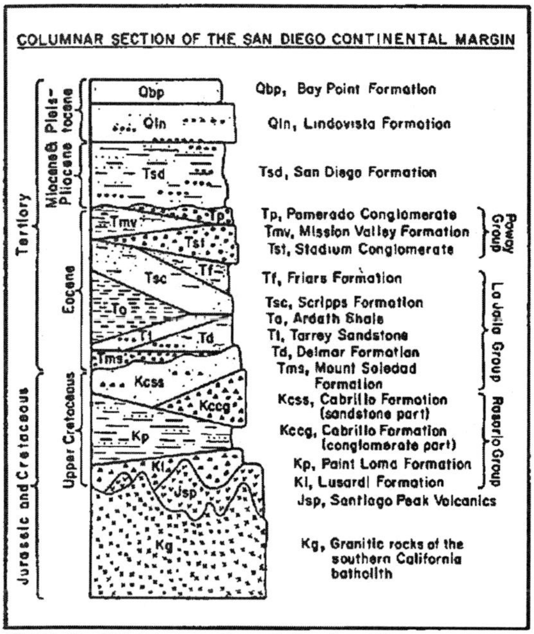

Figure 2. Columnar section of the San Diego continental margin.

Plutonic Rocks of the Southern California Batholith

Plutonic rocks of the southern California batholith in the area are quartz diorite and gabbro. The quartz diorite is typically coarse grained, light gray and contains large phenocrysts of plagioclase and potassium feldspar. Hornblende and biotite are present in small amounts. The gabbro varies considerably in texture and composition but is mostly medium to coarse grained and medium to dark gray. The chief minerals are calcic feldspar and pyroxene, and the accessory minerals include trace amounts of quartz and biotite.

Potassium-argon dates of a gabbro near San Marcos and a quartz diorite 10 km southeast of Escondido are respectively 101 and 105 million years (Evernden and Kistler, 1970). A lead-alpha date on zircon from quartz diorite in the Woodson Mountain area, 20 km southeast of Escondido, is 105 ± 10 million years (Bushee *et al.*, 1963).

Throughout most of the area, the granitic rocks are deeply weathered. Spheroidal boulders, formed as a result of the weathering, range in size from 0.5 meter to 10 meters. The batholithic rocks where weathered are locally very difficult to distinguish from the overlying Eocene Friars Formation, which is largely composed of debris derived from the weathered plutonic basement rock. Careful examination for relict primary features in the plutonic rocks or sedimentary structures in the overlying rocks is necessary to distinguish the weathered basement rock from the sedimentary strata.

Rosario Group

The Rosario Group consists of marine and nonmarine clastic rocks that, oldest to youngest, include the Lusardi Formation, the Point Loma Formation, and the Cabrillo Formation. Only rocks of the Lusardi Formation crop out in the La Mesa-Poway-Escondido area.

Lusardi Formation

The Lusardi Formation, in its type area near Rancho Santa Fe, is a very poorly sorted, deeply weathered boulder conglomerate (Nordstrom, 1970). The few exposures of the Lusardi Formation that occur in the mapped area are in a narrow belt that extends northeastward from the city of Poway (plate 2B). These deposits fill a former stream channel, but the present topography is reversed, with the conglomerate now capping a long narrow ridge. The modern drainage is deeply incised into the granitic rocks on either side of the ridge.

The Lusardi Formation at this locality consists of poorly sorted, angular to well-rounded clasts that range in size from granules to boulders; some of the boulders exceed 3 m in diameter. The matrix is a medium- to fine-grained quartz and feldspar-rich sandstone that comprises about 50 percent of the unit.

The largest and most abundant clasts include coarse-grained diorite, quartz diorite, and medium-grained granodiorite. These rock types, together with minor amounts of aplite and vein quartz, constitute about 60 percent of the clasts in the Lusardi Formation. Other clasts include a variety of fine- to very fine-grained, greenish-gray, and dark-gray metamorphosed tuff. Some of the most distinctive and abundant of these clasts have finely crenulated flow banding on weathered surfaces and are very fine grained, dark, and structureless on fresh surfaces. Less distinctive but abundant clasts are fine-grained black hornfels and volcanic rocks.

Most of the rock types found in the Lusardi Formation are common to the coarse fraction of the Rosario Group as a whole and are considered to be derived largely from local plutonic and metamorphic rocks (Peterson, 1971).

The Lusardi Formation rests unconformably on granitic rocks of the southern California batholith and is in turn overlain by the Eocene Stadium Conglomerate. The character and distribution of the deposits suggest that the clasts of the Lusardi Formation originated east of the area mapped and flowed through a long, narrow, fairly steep-walled river channel in the vicinity of Poway. The Lusardi Formation lies buried below the San Diego coastal area from the vicinity of Carlsbad south to the Mexican Border (Kennedy and Moore, 1971).

EOCENE DEPOSITS

La Jolla Group

The La Jolla Group (Eocene) is composed of intertongued marine, lagoonal, and nonmarine siltstone, sandstone, and conglomerate. These rocks, though partially age equivalent, are from oldest to youngest the Mount Soledad Formation, Del Mar Formation, Torrey Sandstone, Ardath Shale, Scripps Formation, and Friars Formation (Kennedy and Moore, 1971). Only the Friars Formation crops out in the La Mesa-Poway-Escondido area.

Friars Formation

The Friars Formation is a nonmarine and lagoonal sandstone named for exposures along the north side of Mission Valley near Friars Road in the La Jolla quadrangle (Kennedy and Moore, 1971). A molluscan fauna collected from the type section includes *Nekewis io* (Gabb) and *Ectinochilus canalifer supraplicatus* (Gabb). These species together are indicative of the west coast Californian molluscan "Transition stage" and a late middle Eocene age (Givens, 1974).

Most of the area is underlain by the nonmarine facies, which reaches a maximum thickness of 150 m and consists of sandstone with interbeds of claystone. The sandstone is massive, yellowish gray, medium grained, poorly indurated, and caliche-rich. The claystone is dark greenish gray, well indurated, and expansible. Fluviatile cobble conglomerate lenses and tongues that thicken markedly to the east are especially characteristic of the exposures along the eastern margin of the area.

Throughout the mapped area, the Friars Formation rests unconformably on the basement complex and is overlain by sedimentary deposits of Eocene, Pleistocene, and Holocene age.

Landslides are common in the clay-rich part of the formation. The clay is predominantly montmorillonite, but kaolinite is also present. Sixteen clay samples were collected and physical characteristics were analyzed. The results are presented on table 1 and are discussed below under *Landslide Deposits.*

Poway Group

The Poway Conglomerate of Ellis (1919) is one of the most widespread and distinctive rock units in southern California. It crops out primarily in the eastern part of the San Diego area and is the dominant formation in the Poway and La Mesa quadrangles (plates 2B,3B). The rock is mostly nonmarine sandstone and coarse cobble conglomerate composed largely of clasts that have been described in detail by Bellemin and Merriam (1958), DeLisle *et al.* (1965), Woodford *et al.* (1968), and Peterson (1970a).

In a recent revision of the Eocene stratigraphic nomenclature of the San Diego area (Kennedy and Moore, 1971), the Poway Conglomerate was raised to the Poway Group and three formations were recognized within it: a lower conglomerate designated the Stadium Conglomerate, an intermediate sandstone designated the Mission Valley Formation, and an unnamed upper conglomerate formation. This upper conglomerate unit has been subsequently designated the Pomerado Conglomerate (Peterson and Kennedy, 1974).

The arrangement of the three formations in the Poway Group is schematically illustrated in figure 3. The Mission Valley Formation intertongues with the underlying Stadium Conglomerate. Where the Mission Valley Formation pinches out and the Pomerado Conglomerate overlies the Stadium Conglomerate in the eastern part of the area, the two units cannot be distinguished. On the geologic map, where this situation exists, a dashed contact line indicates the approximate location of the boundary between these formations.

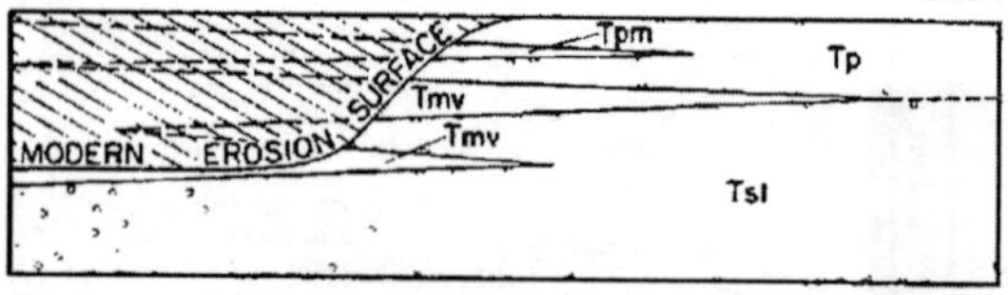

Figure 3. Schematic diagram of lithostratigraphic variations in the Poway Group and modern erosion surface.

Stadium Conglomerate

The type section of the Stadium Conglomerate lies within the SW 1/4 La Mesa quadrangle and is approximately 1 km west of Murphy Canyon Road along the northern wall of Mission Valley near San Diego Stadium (Kennedy and Moore, 1971). At the type section it consists of a cobble conglomerate with a dark yellowish-brown coarse-grained sandstone matrix. The massive conglomerate contains dispersed lenses of fossiliferous crossbedded sandstone. The fossils include calcareous nannoplankton, mollusks, and Foraminifera. The nannoplankton include *Reticulofenestra umbilica* (Levin) and *Discoaster distinctus* Martini. These species when considered with the entire flora collected indicate a middle or late Eocene age (Kennedy, 1973).

The Stadium Conglomerate is moderately well sorted with clasts of cobble size predominating. Boulders as large as 0.5 m in diameter do occur but are extremely rare. Fine-grained rocks within the Stadium Conglomerate generally constitute less than 20 percent of the unit, but locally sandstone beds and lenses may comprise as much as 50 percent of the unit.

The highly distinctive "Poway" clasts consist predominantly (up to 80 percent) of mildly metamorphosed rhyolitic to dacitic volcanic and volcaniclastic rocks and up to 10 percent quartzite. This suite of clasts first appears in the Eocene formations of the San Diego area and is typical of stratal units such as the Stadium Conglomerate and Pomerado Conglomerate. The clasts also are abundant and characteristic in later stratal units such as the Pliocene San Diego Formation, the Pleistocene Lindavista Formation, and the various Quaternary surficial deposits.

The volcanic and pyroclastic clasts of the Poway suite are distinctively different from the local

Santiago Peak Volcanics and no local quartzite outcrops, which compare to the quartzite clasts in the Eocene conglomerate, are known. The provenance of the "Poway" clasts has provoked considerable controversy, and widely differing source areas ranging from the Mojave Desert to Sonora, Mexico, have been proposed (DeLisle *et al.*, 1965; Merriam, 1968; Woodford *et al.*, 1968; Minch, 1972). The direction of stratigraphic thinning, cobble imbrications, and cross-bedding within the Stadium Conglomerate imply that the clasts were transported into their present position from an easterly direction.

The Stadium Conglomerate conformably overlies the Friars Formation and is conformably overlain by the Mission Valley Formation.

Mission Valley Formation

The Mission Valley Formation is composed of marine, lagoonal, and nonmarine sandstone that lies conformably upon the Stadium Conglomerate and is conformably overlain by the Pomerado Conglomerate. The Mission Valley Formation has a maximum thickness of 60 m and was named for exposures along the south wall of Mission Valley on the west side of State Highway 163 in the adjacent La Jolla quadrangle (Kennedy and Moore, 1971). The sandstone is characteristically soft and friable, light olive gray, and fine to medium grained. It is locally interstratified with carbonate cemented beds. Cobble conglomerate tongues within the Mission Valley Formation, which are identical to the Stadium Conglomerate in lithology, comprise up to 30 percent of sections measured in the easternmost exposures but less than 15 percent of sections measured in the western part of the area.

Due to the friable nature of the Mission Valley Formation, it lacks the bold topographic expression of the resistant conglomerate formations that lie stratigraphically above and below. Thin deposits of conglomeratic slope wash commonly mask the Mission Valley Formation in the eastern part of the area, where it is overlain by the Pomerado Conglomerate. The slopes developed in this area are relatively steep on the Pomerado Conglomerate, shallow on the Mission Valley Formation, and steep on the Stadium Conglomerate. An understanding of these topographic relationships helps to determine the distribution of the Mission Valley Formation in areas where it is covered by surficial deposits.

The Mission Valley Formation thins from west to east (figure 3), pinching out in the eastern part of the Poway and La Mesa quadrangles (plates 2B, 3B). The rock contains an upper Eocene molluscan fauna. An assemblage collected from the uppermost beds of the Mission Valley Formation in a road cut 200 m due east of the Miramar Reservoir filtration plant (elevation 238 m) at Lat 32° 54.8′ N.; Long 117° 05.7′ W. includes *Tellina tehachapii* Anderson and Hanna, *Macrocallista Andersoni* Dickerson, *Crassatella uvasana* s.s. Gabb, and *Turritella uvasana sargeanti* (Anderson and Hanna). These species when considered together are indicative of the upper Eocene age (Tejon Stage) and correlative with the upper Eocene of Europe (Givens, 1974).

Pomerado Conglomerate

The Pomerado Conglomerate is the uppermost formation of the Poway Group and has a maximum thickness of 55 meters. It was named for exposures located at the divide between Carroll Canyon and Poway Valley along Pomerado Road (Peterson and Kennedy, 1974). The Pomerado Conglomerate is a massive cobble conglomerate, lithologically identical to the Stadium Conglomerate. The contact between the Mission Valley Formation and Pomerado Conglomerate is conformable and gradational. East of the pinch-out of the Mission Valley Formation, where the Pomerado Conglomerate rests directly on the Stadium Conglomerate, the contact is based on an eastern projection of the feather edge of the Mission Valley Formation along an assumed horizontal surface.

Both the Stadium Conglomerate and Pomerado Conglomerate are characterized by occasional thin beds, lenses, and tongues of light brown medium-grained sandstone. Most of these are not large enough to map. Locally they constitute up to about 20 percent of the formation.

A 10 m thick sandstone lens, designated the Miramar Sandstone Member of the Pomerado Conglomerate (Peterson and Kennedy, 1974), crops out in the vicinity of Miramar Reservoir. Lithologically, the Miramar Sandstone Member is nearly identical to the Mission Valley Formation but is stratigraphically higher and wholly contained within the Pomerado Conglomerate. Its outcropping characteristics and topographic expression are also very similar to those of the Mission Valley Formation.

The Pomerado Conglomerate and associated Miramar Sandstone Member have not yielded fossils; but, on the basis of its stratigraphic relationship with the underlying fossiliferous Mission Valley Formation (figure 3), it is assigned to the upper Eocene.

POST-EOCENE DEPOSITS

Pliocene and Pleistocene Rocks

The Pliocene and Pleistocene rocks include marine sandstone and conglomerate of the Pliocene San Diego Formation, marine and nonmarine sandstone of the late Pliocene or early Pleistocene Lindavista Formation, and lagoonal and nonmarine sandstone of the late Pleistocene Bay Point Formation. The Bay Point Formation is not present in the La Mesa-Poway-Escondido quadrangles.

San Diego Formation

The San Diego Formation (Dall, 1898), middle or late Pliocene in age, crops out along the upper part of the north facing slopes of Mission Valley. These exposures, which attain a maximum thickness of 30 m, are typically yellowish-brown, fine- to medium-grained, poorly indurated sandstone. Cobble conglomerate beds, bentonite, marl, and brown mudstone further characterize the section. The San Diego Formation increases to the south, where it has a maximum thickness of about 400 m (Hertlein and Grant, 1939). The lower 200 m of this section correlates with the Miocene-Pliocene Rosarito Beach Formation in northern Baja California. The cobble-conglomerate stringers are composed primarily of "Poway-type" clasts; but, in some beds, clasts of granitic and metavolcanic rocks, derived from the local basement, comprise up to 50 percent of the total. The bentonite is light brown, waxy to earthy, expansible, and soft.

The San Diego Formation rests unconformably on rocks of the Poway Group and is overlain by the Lindavista Formation. Locally it is separated from the Lindavista Formation by an unconformity, but elsewhere the bedding of the two units is parallel and appears gradational.

Lindavista Formation

The Lindavista Formation was named by Hanna (1926) for exposures at the Lindavista railroad siding 4 km west of the mapped area within the La Jolla 7.5 minute quadrangle. The formation consists of nearshore marine, beach, and nonmarine sediments deposited on a 10 km wide wave-cut platform (Lindavista Terrace of Hanna, 1926) during a period of time that post-dates the San Diego Formation of middle or late Pliocene age and pre-dates the fossiliferous late Pleistocene (Sangamon Stage) Bay Point Formation. A fossil molluscan fauna found in the Lindavista Formation near Lat. 32° 48.5' N.; Long. 117° 6.25' W., includes the extinct species *Pecten bellus*. Because this species is not known from the late Pleistocene, the Lindavista Formation at this locality is considered to be early Pleistocene in age (G. Kennedy, 1973).

The Lindavista Formation in the mapped area is reddish-brown sandstone and conglomerate. Ferruginous cement, mainly hematite, gives the Lindavista Formation its characteristic color and a resistant, ledgy nature.

Both the coarse and fine-grained rocks of the Lindavista Formation have been largely derived from Eocene formations of the area, particularly the Poway Group. Iron-staining is common to the Lindavista Formation, and, where it extends downward into the underlying Eocene rocks, the two become difficult to differentiate. A particularly difficult area for separating the Lindavista Formation from extensively stained Stadium Conglomerate lies east and southeast of Miramar Naval Air Station in the vicinity of Camp Elliott. The upper surface of the Lindavista Formation is commonly characterized by "mima mounds" or "Prairie Mounds," small mound-like hills up to about 10 m in diameter and 1 m high which are useful in differentiating this unit from the rocks of the Poway Group.

Pleistocene and Holocene Surficial Deposits

The Pleistocene and Holocene surficial deposits include stream-terrace, landslide, alluvium, and slope wash deposits.

Stream-Terrace Deposits

Stream-terrace deposits have been preserved in only a few places in the mapped area. These include a poorly consolidated, conglomeratic sand deposit near the confluence of Sycamore Canyon and the San Diego River channel, approximately 2 km west of Santee, and a coarse-grained sand deposit at the mouth of Mission Gorge near Mission Valley. Also unmapped conglomeratic stream-terrace deposits are found in several road cuts excavated for the old Mission Gorge highway, approximately 0.5 km northeast of the gaging station shown on plate 3B.

Landslide Deposits

The area is underlain in large part by incompetent sedimentary rocks which have been broadly dissected by shallow westward-flowing streams. Most of the landslides in the map area are rotational slumps and have occurred along valley walls where rocks of the Friars and Mission Valley Formations occur. The sliding, commonly associated with soft, expansible clay beds within these units, is the result of the combined factors of incompetent rock, ground water, steep slope angle, and basal undercutting of slopes by streams.

Most of the stream channels that dissect the soft sedimentary cover are strongly asymmetrical with their steep side exposed to the north. These slopes are commonly 10 to 15 degrees steeper than those facing south which seldom reach angles greater than 30 degrees. The over-steepening is controlled in part by the presence of resistant impermeable rock layers (Pomerado and Stadium Conglomerate) exposed along the upper slopes as erosional ledges and platforms. The ledges protect the softer incompetent material directly beneath them (Friars and Mission Valley Formations) from erosion. Westward-thinning conglomerate tongues of the Pomerado and Stadium Conglomerates crop out along the upper valley slopes over a large part of the area that lies between Rancho Bernardo and Fortuna Mountain (plates 2B, 3B). Landslides occur beneath these beds in the soft sandstone and claystone of the Friars and Mission Valley Formations.

Several man-induced slides in the Rancho Bernardo area were studied, and all were found to occur beneath resistant conglomerate layers within the soft sandstone and claystone. Stability filling (compacted fill placed over a benched cut slope) may be one means by which this type of failure can be avoided. Because landslide incidence is greatly increased during periods of high rainfall, as a result of lowered internal rock strength, subdrainage may be another means of slope control.

Slopes steeper than 30 degrees underlain by clay-rich facies of the Friars Formation in the Rancho Bernardo, Poway Valley, and Mission Gorge areas are mantled with surficial landslide debris that coalesces with slope wash and alluvium in the valley bottoms. Sixteen samples of claystone were collected from these areas (plates 1B-3B) and analyzed for their physical properties.

The results of Atterberg tests and the quantity of individual size fractions from particle size distribution tests of these samples are shown in table 1.

For engineering purposes the clay fraction is defined as less than 0.002 millimeter (mm), the silt fraction from 0.002 to 0.074 mm, and sand from 0.074 to 2.0 millimeters. The liquid limit (LL) is defined as the minimum moisture content at which the material behaves as liquid using this test. The plastic limit (LP) is defined as the minimum moisture content at which the sample behaves plastically using this specified test procedure. The plasticity index (IP) is the numerical difference between the liquid limit and the plastic limit. The plasticity index is then the range of moisture content over which the sample behaves plastically. There exists a direct relationship between the liquid limit and "compression index," and between the liquid limit and the "coefficient of consolidation," whereas an inverse relationship exists between the plasticity index and "shearing resistance" (Terzaghi and Peck, 1967).

The stated median particle size was taken from the 50 percent cumulative level of particle size distribution graphs. The arithmetic mean of the median particle sizes among all the samples is 0.016 mm (16 microns), which is in the lower portion of the silt range.

Associations between plasticity index, particle size, and the percentage of clay-size particles are readily observed from the data of table 1. The medium-plasticity samples (IP = 10 to 20) have an average median particle size of 47 microns. High-plasticity samples (IP = 20 to 40) have an average median particle size of 14 microns. The very-high-plasticity samples (IP > 40) have a mean particle size of 6 microns. A direct relationship between higher plasticity and landslide incidence can be seen by comparing this data with field observations in that both the surficial and bedrock landslides in the areas sampled are more abundant with an increase in the plasticity index. Nearly all of the landslides mapped have occurred in rocks with a plasticity index greater than 20.

Alluvium and Slope Wash

Alluvium in the area consists primarily of poorly consolidated stream deposits of silt, sand, and cobble-sized particles derived from bedrock sources that lie within and to the east of the study area. The alluvium is intertongued with Holocene slope wash that generally mantles the lower valley slopes throughout the area. For this reason, alluvium and slope wash have not been differentiated in most areas.

The slope wash deposits consist primarily of poorly consolidated surficial materials derived from nearby soil and decomposed bedrock sources. This reworked debris is deposited along the flanks of the lower valley slopes by the action of gravity and surface water. Thick deposits of slope wash are commonly associated with thick soil horizons developed on the Friars and Mission Valley Formations. Expansive clay horizons weathered from bedrock sources and deposited as slope wash yield the hummocky topography that is common to much of this area.

STRUCTURE AND SEISMIC HISTORY

The oldest rocks in the study area, the upper Jurassic Santiago Peak Volcanics are massive, complexly deformed, and their structure within the mapped area is not readily decipherable. They have undergone low-grade metamorphism and have been intruded by rocks of the mid-Cretaceous southern California batholith.

Regional uplift followed deformation, metamorphism, and batholithic intrusion near the close of the Mesozoic Era, and deep-seated batholithic rocks were extensively exposed. An erosion surface having in excess of 500 m relief was developed on these rocks, setting the stage for deposition of sedimentary rocks in the Late Cretaceous and Tertiary periods (Peterson and Nordstrom, 1970).

The basement rocks have acted as a rigid platform from Late Cretaceous time to the present and the post-batholithic sedimentary rocks deposited upon them are only slightly deformed and mostly flat-laying (section A-A', B-B'). Mapping of the rock units over a broad area has demonstrated that inclinations locally associated with Tertiary and Quaternary faulting are rarely greater than 2 degrees.

Evidence for Late Cenozoic uplift and faulting within the mapped area is abundant (Moore and Kennedy, 1970; Peterson, 1970b; Moore, 1972; Ziony and Buchanan, 1972). Uplift of the Lindavista terrace is evident in that the early Pleistocene shoreline associated with the present landward extension of the Lindavista Formation lies at an altitude of nearly 165 m in the western part of the Poway and La Mesa quadrangles.

The Poway terrace, which is extensively developed in the eastern part of the Poway and La Mesa quadrangles, lies at an altitude of about 275 to 325 meters. Hanna (1926) and others consider the Poway terrace to be the result of Pleistocene marine planation like that of the Lindavista terrace. Possibly this planar surface is a stripped structural surface developed on the resistant upper surface of the Pomerado Conglomerate.

Pleistocene or younger faults in the study area occur in the vicinity of Collwood Boulevard and Montezuma Road, Murphy Canyon, and Mission Gorge. A post-Lindavista fault is inferred to coincide with at least the southern part of Murphy Canyon and the southern part of Mission Gorge because the Lindavista Formation lies topographically higher on the west side of these canyons.

Holocene seismic activity along several faults that lie within 10 km of the area is supported by (1) the historic seismicity believed to be associated with the Rose Canyon fault zone in the San Diego Bay area and (2) subbottom acoustic profiles showing Holocene sediments offset on the sea floor at a location 25 km north of La Jolla within the Rose Canyon fault zone (Moore, 1972).

Table 1. Atterberg limits and particle size distribution.

No. on map	Location (All San Bernardino Base and Meridian)	Atterberg Limits			Particle Size			Mineralogy*
		Liquid limit	Plastic limit	Plasticity index	Sand (%)	Silt (%)	Clay (%)	
1	Sec. 31, T. 13 S., R. 2 W., 1350 m south of intersection of Black Mountain Road and Rancho Bernardo Road in small road cut on east side.	55	30	25	19	69	12	Mostly montmorillonite, some mica, trace of kaolinite.
2	Sec. 22, T. 13 S., R. 2 W., road cut on southwest corner of intersection of Rancho Bernardo and West Bernardo Drive, 3 m above road level.	52	22	30	13	83	4	Mostly montmorillonite, some mica, trace of kaolinite, and minor quartz.
3	Sec. 27, T. 13 S., R. 2 W., Rancho Bernardo Industrial Center development. Cut beneath southeast corner of National Cash Register Co. building, 3 m below contact between the Stadium Conglomerate and Friars Formation.	79	30	49	8	60	32	Mostly montmorillonite, trace of kaolinite.
4	Sec. 27, T. 13 S., R. 2 W., 40 m west of intersection of Center Drive and U.S. Highway 395 near Rancho Bernardo in small cut on south side at base of exposure.	58	27	31	3	61	36	Mostly montmorillonite, trace of quartz.
5	Sec. 27, T. 13 S., R. 2 W., 275 m south of Lomica Drive in Rancho Bernardo on Center Drive in high road cut on east side at elevation 180 m (approximately 1.5 m above street level).	53	22	31	8	61	31	Mostly montmorillonite, trace of kaolinite and quartz.
6	10 m vertically above number 5.	44	19	25	21	55	24	Mostly montmorillonite, trace of quartz.
7	22 m vertically above number 5.	74	29	45	0	60	40	Mostly montmorillonite, trace of quartz.
8	Sec. 17, T. 14 S., R. 2 W., 850 m north of Poway Road on east side of U.S. Highway 395 in road cut. Sample collected 3 m above road level at contact between green claystone of the Friars Formation and rocks of the Santiago Peak Volcanics.	63	25	38	14	69	17	Mostly montmorillonite, minor quartz.
9	3 m vertically above number 8 at contact between Stadium Conglomerate and Friars Formation.	60	20	40	4	62	34	Mostly montmorillonite, minor quartz.
10	Sec. 14, T. 14 S., R. 2 W., 1125 m east of Pomerado Road on north side of Poway Road in road cut at base.	47	25	22	27	56	17	Mostly montmorillonite, minor quartz, trace feldspar.
11	Sec. 26, T. 14 S., R. 2 W., 600 m south of intersection of Pomerado Road near Beeler Canyon Road on Pomerado Road; 2 m beneath Stadium Conglomerate on west side of road.	57	21	36	4	82	14	Mostly montmorillonite, trace of mica, trace of kaolinite, minor quartz and feldspar, trace of calcite.
12	Sec. 35, T. 15 S., R. 2 W., intersection of Old Mission Gorge Road and New Mission Gorge Road at southeast corner 900 m east of number 13 and 1 m above road level.	60	10	50	15	55	30	Mostly montmorillonite, trace of mica, minor kaolinite, trace of quartz and feldspar.

continued on page following....

Table 1. Atterberg limits and particle size distribution (continued).

No. on map	Location (All San Bernardino Base and Meridian)	Atterberg Limits			Particle Size			Mineralogy*
		Liquid limit	Plastic limit	Plasticity index	Sand (%)	Silt (%)	Clay (%)	
13	Sec. 2, T. 16 S., R. 2 W., southeast side of Mission Gorge Road 900 m south of intersection of Old Mission Gorge Road and 1140 m northeast of Conestoga Way. Locality is 1 m above road level.	66	25	41	24	47	29	Mostly montmorillonite, minor kaolinite, trace of quartz and feldspar.
14	Sec. 18, T. 18 S., R. 3 W., 400 m west of Murphy Canyon Road on Friars Road in high cut on north wall of Mission Valley, 3 m above base of cut and 5 m beneath Stadium Conglomerate.	52	21	31	12	63	25	Mostly montmorillonite, trace of mica, minor kaolinite, trace of quartz and feldspar.
15	Sec. 16, T. 18 S., R 3 W., 365 m northeast of intersection between Interstate Highway 8 and Waring Road; 2 m above base of cut on Frontage Road, 244 m east of Waring Road.	59	24	35	38	46	16	Mostly montmorillonite, trace of mica, minor kaolinite, trace of quartz, feldspar and calcite.
16	Sec. 15, T. 18 S., R. 3 W., 550 m due west of the intersection between Interstate Highway 8 and College Avenue on south side of Mission Valley in road cut at San Diego State College, 3 m above road level.	52	20	32	5	74	21	Mostly montmorillonite, minor mica and quartz, trace of feldspar and calcite.

* Analyses made by Paul Anderson, California Division of Mines and Geology Laboratory.

Forty-four earthquakes with magnitudes between 2.5 and 3.7 have been recorded within the greater San Diego metropolitan area since 1950. Three of these which occurred in 1964 on June 21, June 22, and July 14 had epicenters within the vicinity of San Diego Bay and magnitudes of 3.7, 3.6, and 3.5 respectively.

The San Diego area has experienced ground shaking produced by earthquakes with epicenters as distant as 100 kilometers. The 1933 Long Beach earthquake (M 6.6) caused minor damage in San Diego County and was felt sharply as far south as the Mexican border.

On November 4, 1949, and February 9, 1956, earthquakes on the Vallecitos-San Miguel fault in northern Baja California were felt in San Diego. The 1949 earthquake (M 5.7) had an epicentral distance of approximately 75 km southeast of San Diego. The epicenter of the 1956 earthquake (M 6.8) was approximately 175 km south of San Diego and caused ground rupture for 20 km in the vicinity of the epicenter (Shor and Roberts, 1958). Three aftershocks of this earthquake with magnitudes greater than 6 occurred on February 9, 14, and 15, 1956. Several earthquakes (M > 5) on the Agua Blanca fault, also in northern Baja California, have been recorded during the past 30 years with epicentral distances within 125 km of the San Diego civic center.

During the past 35 years at least ten of the strongest earthquakes that have occurred in the Imperial Valley-Salton Trough have been felt in San Diego. Three of these, the 1940 Imperial Valley earthquake (M 7.1), the 1951 Superstition Hills earthquake (M 5.6), and the 1968 Borrego Mountain earthquake (M 6.5) caused minor damage in western San Diego County.

The February 9, 1971, San Fernando Valley earthquake (M 6.4) was felt sharply in San Diego. Minor damage occurred in the coastal area as far south as National City. Several small rockfalls occurred in roadcuts on Highway 395 between Poway and Miramar Roads.

MINERAL RESOURCES

Mineral resources in the Escondido, Poway, and La Mesa quadrangles (plates 1B-3B) include extensive deposits of sand, gravel, and metavolcanic rock suitable for use as aggregate in highway asphalt, portland cement, and ceramic products. Large reserves of decomposed granite, small tonnages of pyrophyllite, and minor amounts of arsenopyrite, gold, silver, and uranium are also present.

Clay in the Friars Formation has not been commercially mined within the mapped area but represents a potential source of expansible clay, fire clay, and lightweight aggregate. Table 1 summarizes the physical properties of 16 clay samples collected from the Friars Formation.

The mineral resources and mineral industry of San Diego County are discussed in detail by Weber (1963). The mines, pits, and quarries in the area are listed in table 2. Mineral-resource inventories are made annually by the Natural Resources Division, San Diego County Department of Agriculture, and are available to the public through the agency.

Table 2. Mines, quarries, and pits in the La Mesa, Poway, and SE¼ Escondido quadrangles.

*Mines, quarries, or pits**	*Location (All San Bernardino Base and Meridian)*	*Geologic unit*	*Mineral resource*	*Remarks*
1. Bly Stone Co. Quarry.	Sec. 10, T. 13 S., R 2 W.	Plutonic rock of the southern California batholith.	Dimension stone.	Operated from 1921 to 1924 for large unfractured blocks of San Marcos Gabbro.
2. Van Deventer quarry (Daley Corporation).	Sec. 10 (?) T. 13 S., R 2 W.	Plutonic rock of the southern California batholith.	Dimension stone.	Reported by Tucker (1925) to be on the south shore of Lake Hodges but exact location is undetermined.
3. Four-Gee deposit (Golem).	Sec. 19, T. 13 S., R 2 W.	Santiago Peak Volcanics.	Pyrophyllite.	Discovered in 1952 and mined since 1953.
4. Property now owned by Rancho Bernardo Inc.	Sec. 27, T. 13 S., R 2 W.	Plutonic rock of the southern California batholith.	Decomposed granite.	Inactive 1972; used for roadbed fill during expansion of U.S. Highway 395.
5. Black Mountain deposit; Oliver Wylie estate.	Sec. 5, T. 14 S., R 2 W.	Santiago Peak Volcanics.	Arsenopyrite (Arsenic) minor amount of gold and silver.	Operated in 1924 for arsenic and gold. Total recovery reported was 700 pounds of material containing 31.4 percent arsenic plus a small amount of gold and silver.
6. C. B. Grove (Pit No. 2).	Sec. 10, T. 14 S., R 2 W.	Plutonic rock of the southern California batholith.	Decomposed granite.	Inactive 1972; active in 1957.
7. Fletcher Quarries.	Sec. 15, T. 14 S., R 2 W.	Plutonic rock of the southern California batholith.	Decomposed granite.	Inactive 1972.
8. Einer Brothers (Poway Pit).	Sec. 14, T. 14 S., R 2 W.	Plutonic rock of the southern California batholith.	Decomposed granite.	Inactive 1972.
9. Candel and Johnson (Poway Operation).	Sec. 21 & 22, T. 14 S., R 2 W.	Stadium Conglomerate.	Concrete sand and crushed gravel.	Inactive 1972; active 1958.
10. Escondido Sand and Gravel Works.	Sec. 26, T. 14 S., R 2 W.	Alluvium derived mostly from rocks of the Poway Group.	Concrete and bituminous aggregate.	Inactive 1958.
11. San Diego Consolidated Co., (Carrol Canyon Plant).	Sec. 6, T. 15 S., R 2 W.	Stadium Conglomerate and Alluvium.	Bituminous aggregate.	Active 1972.
12. Nelson and Sloan (Miramar Plant).	Sec. 19, T. 15 S., R 2 W.	Stadium Conglomerate and Alluvium.	Concrete sand and crushed gravel.	Began operation 1956, active 1972.
13. Fenton, H. G., Material Co.	Sec. 20, T. 15 S., R 1 W.	Alluvium.	Concrete and plaster sand, and crushed gravel.	Inactive 1972.
14. Fenton, H. G., Material Co.	Sec. 25, T. 15 S., R 2 W. & Sec. 30, T. 15 S., R 1 W.	Alluvium.	Concrete and plaster sand, and crushed gravel.	Operation began in 1954 in 200 acres; inactive 1972.
15. Acme Truck Co. (Pit #1)	Sec. 35, T. 15 S., R 2 W.	Plutonic rocks of the southern California batholith.	Decomposed granite.	Inactive 1972.

continued on page following....

Table 2. Mines, quarries, and pits in the La Mesa, Poway, and SE¼ Escondido quadrangles (continued).

*Mines, quarries, or pits**	*Location (All San Bernardino Base and Meridian)*	*Geologic unit*	*Mineral resource*	*Remarks*
16. Fletcher Quarries—Ed Fletcher Co.	Sec. 35, T. 15 S., R 2 W.	Santiago Peak Volcanics.	Riprap.	Used in construction of Mission Bay Park and jetty.
17. Industrial Asphalt (Plant 36).	Sec. 35, T. 15 S., R 2 W.	Santiago Peak Volcanics.	Bituminous aggregate.	Active 1972.
18. Dennis, V. R., Canyon Rock Co.	Sec. 3, T. 16 S., R 2 W.	Santiago Peak Volcanics.	Riprap, concrete aggregate, bituminous aggregate.	Began operations 1929.
19. Dennis, V. R., Canyon Rock Co.	Sec. 3, T. 16 S., R 2 W.	Alluvium.	Ceramic and concrete sand.	San Diego River bed sand operation.
20. Daley Corp.	Secs. 5 & 8, T. 16 S., R 2 W.	Friars Formation and Stadium Conglomerate.	Crushed gravel, for concrete and bituminous aggregate.	Active 1972.
21. Denton, American Sand Inc.	Sec. 8, T. 16 S., R 2 W.	Alluvium.	Plaster sand.	San Diego River bed; began operations in 1951, active 1972.
22. Nelson and Sloan, Mission Sand Co.	Sec. 17, T. 16 S., R 2 W.	Alluvium.	Plaster sand.	San Diego River bed; inactive 1972.
23. Fenton, H. G., Material Co. (Mission Valley Plant).	Sec. 18, T. 16 S., R 2 W.	Poway Group and Alluvium.	Concrete sand and crushed gravel.	Active 1972.
24. Olswick Prospect.	Sec. 32, T. 15 S., R 1 W.	Stadium Conglomerate.	Unidentified uranium minerals.	Mineralized zone in 3m wide bulldozer pit in sandstone bed near contact with underlying granitic basement rock. An unpublished short report on the site was completed by the Atomic Energy Commission in 1955.
25. Independent Stone Co.	Sec. 30, T. 16 S., R 1 W.	Santiago Peak Volcanics.	Crushed stone.	Inactive 1972.

* Compiled in part from Weber, 1963.

REFERENCES CITED

Bellemin, G.J., and Merriam, R., 1958, Petrology and origin of the Poway conglomerate, San Diego County, California: Geol. Soc. America Bull., v. 69, p. 199-220.

Bukry, David, and Kennedy, M.P., 1969, Cretaceous and Eocene coccoliths at San Diego, California, *in* Short contributions to California geology: California Division of Mines and Geology Special Report 100, p. 33-43.

Bushee, J., Holden, J., Geyer, B., and Gastil, G., 1963, Lead-alpha dates for some basement rocks of southwestern California· Geol. Soc. America Bull., v. 74, p. 803-806.

Dall, W.H., 1898, 18th Ann. Rept.: U.S. Geol. Survey, pt. 2, correlation table opp. p. 334.

DeLisle, M., Morgan, J.R., Heldenbrand, J., and Gastil, G., 1965, Lead-alpha ages and possible sources of metavolcanic rock clasts in the Poway conglomerate, southwest California: Geol. Soc. America Bull., v. 76, p. 1069-1074.

Ellis, A.J., 1919, Geology, western part of San Diego county, California: U.S. Geol. Survey Water-Supply Paper 446, p. 50-76.

Evernden, J.F., and Kistler, R.W., 1970, Chronology of emplacement of Mesozoic batholithic complexes in California and western Nevada: U.S. Geol. Survey Prof. Paper 623, 42 p.

Fife, D.L., Minch, J.A., and Crampton, P.J., 1967, Late Jurassic Age of the Santiago Peak Volcanics, California: Geol. Soc. America Bull., v. 78, p. 299-304.

Givens, C.R., 1974, Eocene molluscan biostratigraphy of the Pine Mountain area, Ventura County, California: University of California, Dept. of Geol. Sci. Bull., v. 109, 107 p.

Gray, C.H., Jr., Kennedy, M.P., and Morton, P.K., 1971, Petroleum potential of southern coastal and mountain area, California: American Assoc. Petroleum Geologists, Mem. 15, p. 372-383.

Hanna, M.A., 1926, Geology of the La Jolla quadrangle, California: University of California, Dept. Geol. Sci. Bull., v. 16, p. 187-246.

Hertlein, L.G., and Grant, U.S., IV., 1939, Geology and oil possibilities of southwestern San Diego County: California Journal Mines and Geol., v. 35, p. 57-78.

Hertlein, L.G., and Grant, U.S., IV, 1944, The geology and paleontology of the marine Pliocene of San Diego, California, pt. 1, Geology: San Diego Soc. Nat. History Mem., v. 2, p. 1-72.

Kennedy, G.L., 1973, Early Pleistocene invertebrate faunule from the Lindavista Formation, San Diego, California: San Diego Soc. of Nat. History, Transactions, v. 17, p. 119-128.

Kennedy, M.P., 1973, Stratigraphy of the San Diego embayment, California: Unpublished Ph.D. dissertation, University of California, Riverside.

Kennedy, M.P., 1975, Geology of the Del Mar, La Jolla and Point Loma quadrangles, San Diego metropolitan area, San Diego County, California: California Div. Mines and Geology Bull. 200A.

Kennedy, M.P., and Moore, G.W., 1971, Stratigraphic relations of upper Cretaceous and Eocene Formations, San Diego coastal area, California: American Assoc. Petroleum Geologists Bull., v. 55, p. 709-722.

Kern, J.P., 1971, Paleoenvironmental analysis of a late Pleistocene estuary in southern California: Journal of Paleo., v. 45, p. 810-823.

Larsen, E.S., 1948, Batholith and associated rocks of Corona, Elsinore, and San Luis Rey quadrangles, southern California: Geol. Soc. America Mem. 29, 182 p.

Merriam, R., 1968, Geologic reconnaissance of northwest Sonora: Stanford University Pubs. Geol. Sci., v. 11, p. 287.

Milow, E.D., and Ennis, D.B., 1961, Guide to geologic field trip of southwestern San Diego County: Geol. Soc. America Cordilleran Sec. 57th Ann. Mtg., Guidebook, p. 23-43.

Minch, J.A., 1972, The late Mesozoic—early Tertiary framework of continental sedimentation, northern peninsular ranges, Baja California, Mexico: Unpublished Ph.D. Dissertation, University of California, Riverside.

Moore, G.W., 1972, Offshore extension of the Rose Canyon fault, San Diego, California: U.S. Geol. Survey Prof. Paper 800-C.

Moore, G.W., and Kennedy, M.P., 1970, Coastal geology of the California-Baja California border area: American Assoc. Petroleum Geologists Guidebook, Pacific Section fall field trip, p. 4-9.

Nordstrom, C.E., 1970, Lusardi Formation—a post-batholithic Cretaceous conglomerate north of San Diego, California: Geol. Soc. America Bull., v. 81, p. 601-605.

Peterson, G.L., 1970a, Distinctions between Cretaceous and Eocene conglomerates in the San Diego area, southwestern California: American Assoc. Petroleum Geologists Guidebook, Pacific Section fall field trip, p. 90-98.

Peterson, G.L., 1970b, Quaternary deformation of the San Diego area, southwestern California: American Assoc. Petroleum Geologists Guidebook, Pacific Section fall field trip, p. 120-126.

Peterson, G.L., 1971, Stratigraphy of the Poway area, southwestern California: San Diego Soc. Nat. History Transactions, v. 16, no.

Peterson, G.L., and Kennedy, M.P., 1974, Lithostratigraphic variations in the Poway Group near San Diego, California: San Diego Soc. of Nat. History Transactions, v. 17, p. 251-258.

Shor, G.G., Jr., and Roberts, E., 1958, San Miguel, Baja California Norte, earthquakes of February, 1956; a field report: Seismological Society of America Bull., v. 48, p. 101-116.

Terzaghi, K., and Peck, R.B., 1967, Soil mechanics in engineering practice: John Wiley and Son, New York, 729 p.

Weber, F.H., Jr., 1963, Geology and mineral resources of San Diego County, California: California Div. Mines and Geology County Rept. 3, 309 p.

Woodford, A.O., Welday, E.E., and Merriam, R., 1968, Siliceous tuff clasts in the upper Paleocene of southern California: Geol. Soc. America Bull., v. 79, p. 1461-1486.

Ziony, J.I., and Buchanan, J.M., 1972, Preliminary report on recency of faulting in the greater San Diego area, California: U.S. Geol. Survey Open-File Rept., 16 p.

STATE OF CALIFORNIA
THE RESOURCES AGENCY
DEPARTMENT OF CONSERVATION
CALIFORNIA DIVISION OF MINES AND GEOLOGY
JAMES E. SLOSSON, STATE GEOLOGIST
Solana Beach
Del Mar
SAN DIEGUITO
PACIFIC

PREPARED IN COOPERATION
WITH
THE CITY OF SAN DIEGO

SAN DIEGO METROPOLITAN AREA
BULLETIN 200, PLATE 1A

GEOLOGY OF THE DEL MAR QUADRANGLE
SAN DIEGO COUNTY, CALIFORNIA

by Michael P. Kennedy

SCALE 1:24 000

CONTOUR INTERVAL 20 FEET
DOTTED LINES REPRESENT 10-FOOT CONTOURS
DATUM IS MEAN SEA LEVEL
DEPTH CURVES AND SOUNDINGS IN FEET—DATUM IS MEAN LOWER LOW WATER

1975

EXPLANATION

QUATERNARY

Holocene

Qaf — Artificial fill

Qb — Beach sand

Qal + Qsw — Alluvium and slopewash undifferentiated

Qls — Landslide deposits

Pleistocene

Stream-terrace deposits

Qbp — Bay Point Formation

Qln / Qlb — Lindavista Formation

Qln, nearshore deposits, Qlb, beach deposits

TERTIARY

Eocene

Poway and La Jolla Groups

Tmv, Mission Valley Formation; Tst, Stadium Conglomerate; Tf, Friars Formation; Tscu, Scripps Formation (upper tongue); Tsc, Scripps Formation; Ta, Ardath Shale; Tt, Torrey Sandstone; Td, Delmar Formation; Td & Tf, Delmar and Friars Formation undifferentiated. Conglomerate marked by circle pattern, sandstone marker bed shown by dot pattern.

CRETACEOUS

Upper Cretaceous

Lusardi Formation

Kgb — Gabbro

JURASSIC

Jsp — Santiago Peak Volcanics

SYMBOLS

Contact

O C E A N
TORREY PINES PARK
TORREY PINES PARK LANDSLIDE
SOLEDAD VALLEY
SAN DIEGO
CARMEL VALLEY FAULT
SURFICIAL DEPOSITS NOT SHOWN ON CROSS SECTION
Upper Cretaceous and Eocene rocks Undifferentiated
Pacific Ocean
TOPOGRAPHIC BASE MAP

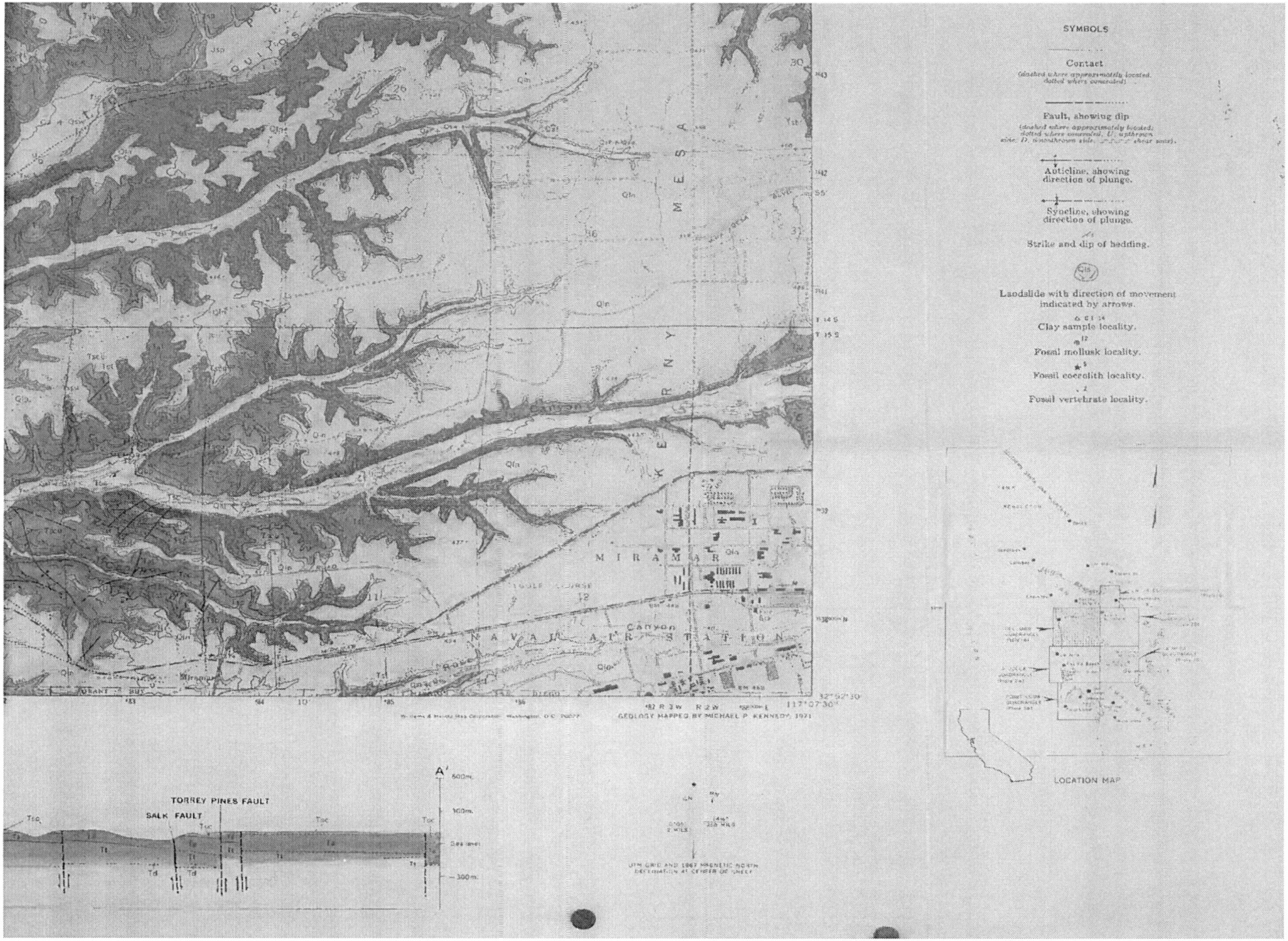
SYMBOLS
Contact
(dashed where approximately located, dotted where concealed)
Fault, showing dip
Anticline, showing direction of plunge.
Syncline, showing direction of plunge.
Strike and dip of bedding.
Landslide with direction of movement indicated by arrows.
Clay sample locality.
Fossil mollusk locality.
Fossil coccolith locality.
Fossil vertebrate locality.
LOCATION MAP
MESA
KEARNY
MIRAMAR
NAVAL AIR STATION
Canyon
GOLF COURSE
GEOLOGY MAPPED BY MICHAEL P. KENNEDY, 1971
R 3 W R 2 W
T 14 S
T 15 S
32°52'30"
117°07'30"
TORREY PINES FAULT
SALK FAULT
A'
600m.
300m.
Sea level
-300m.
UTM GRID AND 1967 MAGNETIC NORTH DECLINATION AT CENTER OF SHEET

CALIFORNIA DIVISION OF MINES AND GEOLOGY
JAMES E. SLOSSON, STATE GEOLOGIST
STATE OF CALIFORNIA
THE RESOURCES AGENCY
DEPARTMENT OF CONSERVATION
PREPARED IN COOP
WITH
THE CITY OF SAN
SAN DIEGO
CORONADO
Harbor
FORT ROSECRANS
LANDSLIDE
Ballast Point
Zuniga Point
FAULT
ZONE

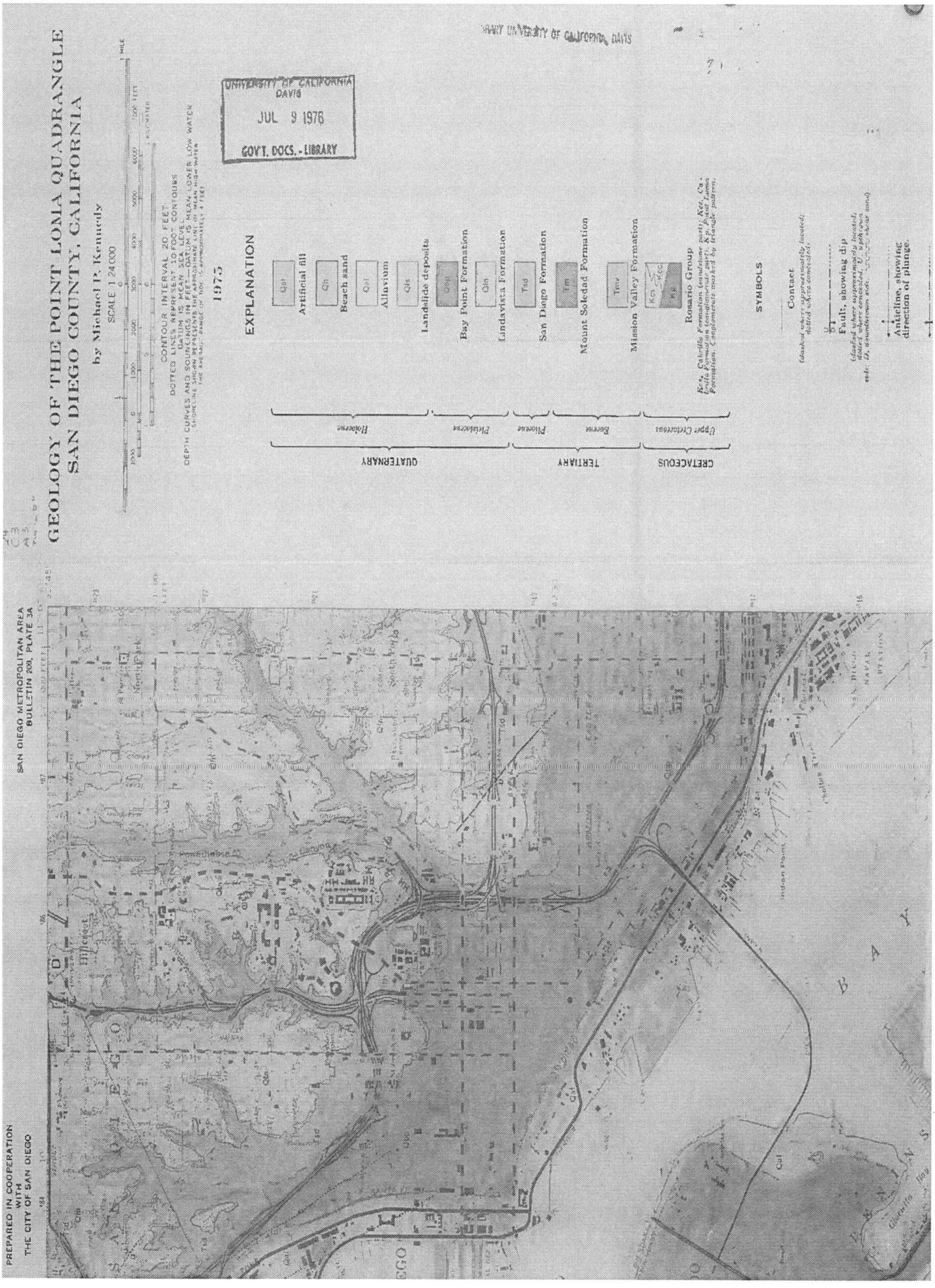
PREPARED IN COOPERATION WITH THE CITY OF SAN DIEGO
SAN DIEGO METROPOLITAN AREA BULLETIN 200, PLATE 3A
GEOLOGY OF THE POINT LOMA QUADRANGLE
SAN DIEGO COUNTY, CALIFORNIA
by Michael P. Kennedy
SCALE 1:24 000
CONTOUR INTERVAL 20 FEET
DOTTED LINES REPRESENT 10 FOOT CONTOURS
DATUM IS MEAN SEA LEVEL
DEPTH CURVES AND SOUNDINGS IN FEET—DATUM IS MEAN LOWER LOW WATER
1975
EXPLANATION
QUATERNARY
Holocene
Pleistocene
TERTIARY
Pliocene
Eocene
CRETACEOUS
Upper Cretaceous
Qaf
Artificial fill
Qb
Beach sand
Qal
Alluvium
Qls
Landslide deposits
Qbp
Bay Point Formation
Qln
Lindavista Formation
Tsd
San Diego Formation
Tm
Mount Soledad Formation
Tmv
Mission Valley Formation
Kcs
Kcc
Kp
Rosario Group
SYMBOLS
Contact
Fault, showing dip
Anticline, showing direction of plunge.
UNIVERSITY OF CALIFORNIA DAVIS
JUL 9 1976
GOV'T. DOCS. - LIBRARY
Hillcrest
North Park
South Park
Indian Point
SAN DIEGO NAVAL STATION
BAY

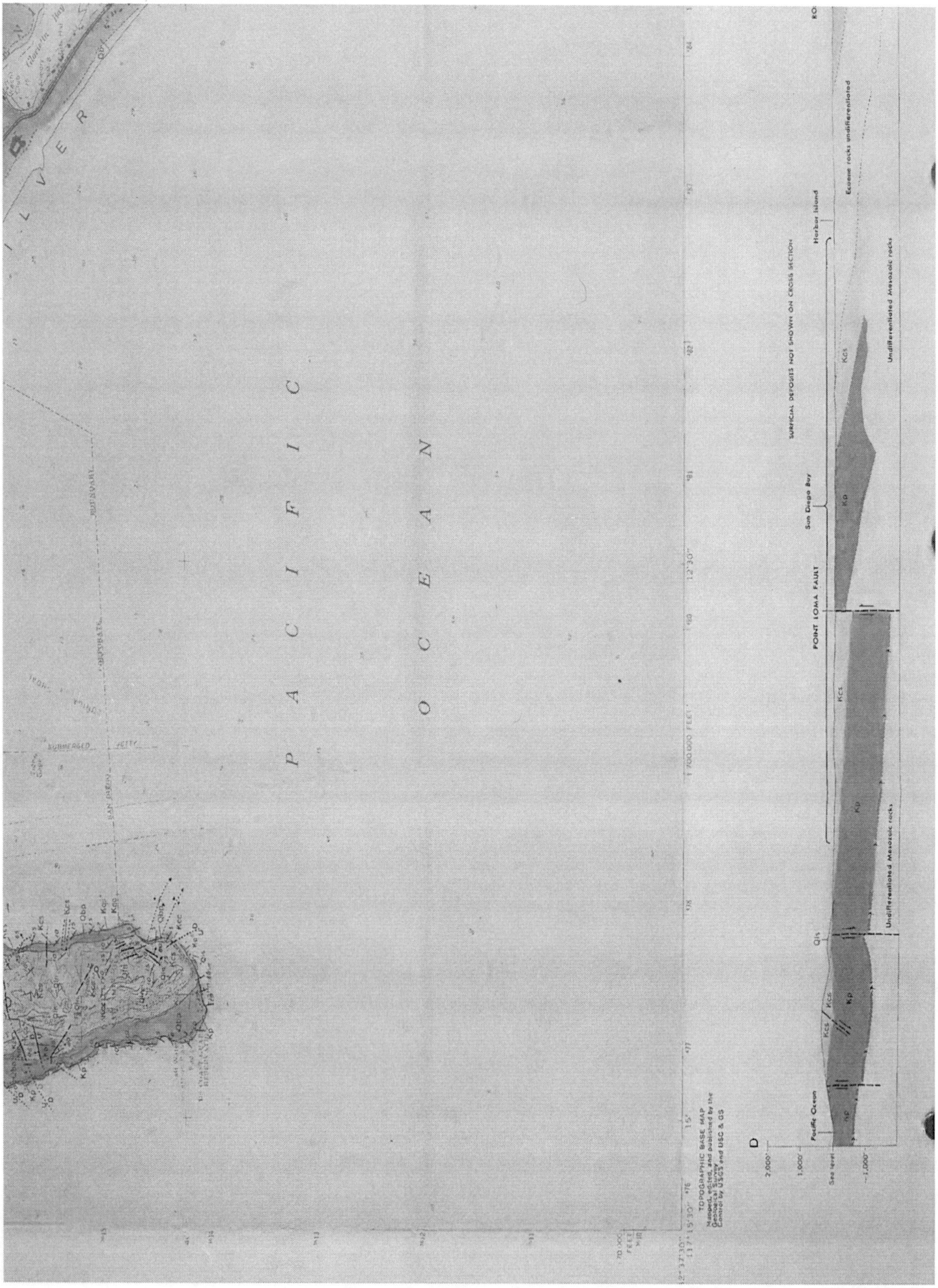

PACIFIC OCEAN
POINT LOMA FAULT
San Diego Bay
Harbor Island
SURFICIAL DEPOSITS NOT SHOWN ON CROSS SECTION
Eocene rocks undifferentiated
Undifferentiated Mesozoic rocks
Pacific Ocean
D
2,000'
1,000'
Sea level
-1,000'
TOPOGRAPHIC BASE MAP
Mapped, edited, and published by the Geological Survey
Control by USGS and USC & GS
1 700 000 FEET

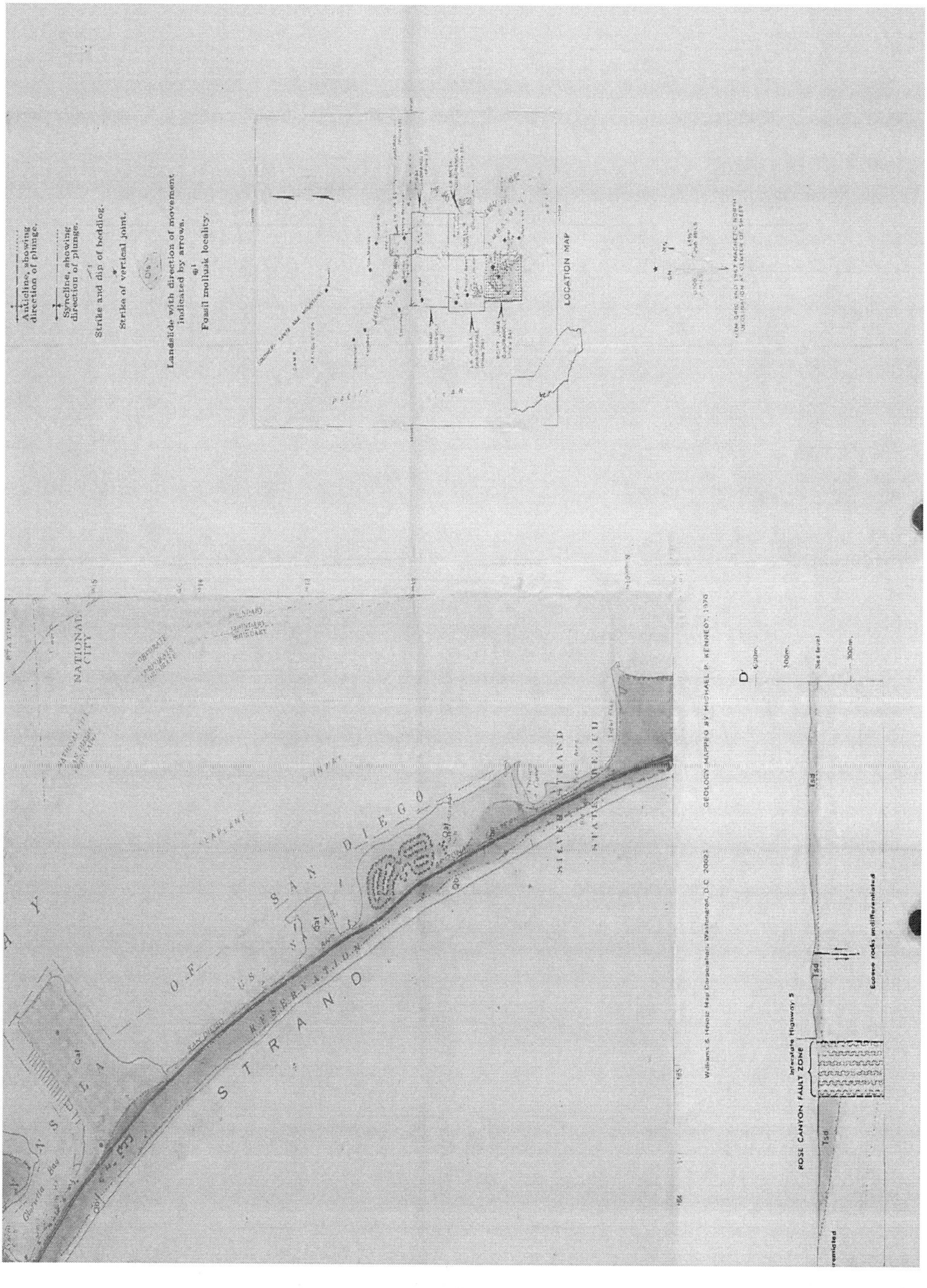
Anticline, showing direction of plunge.
Syncline, showing direction of plunge.
Strike and dip of bedding.
Strike of vertical joint.
Landslide with direction of movement indicated by arrows.
Fossil mollusk locality.
LOCATION MAP
NATIONAL CITY
STRAND
GEOLOGY MAPPED BY MICHAEL P. KENNEDY, 1970
ROSE CANYON FAULT ZONE
Interstate Highway 5
Eocene rocks undifferentiated

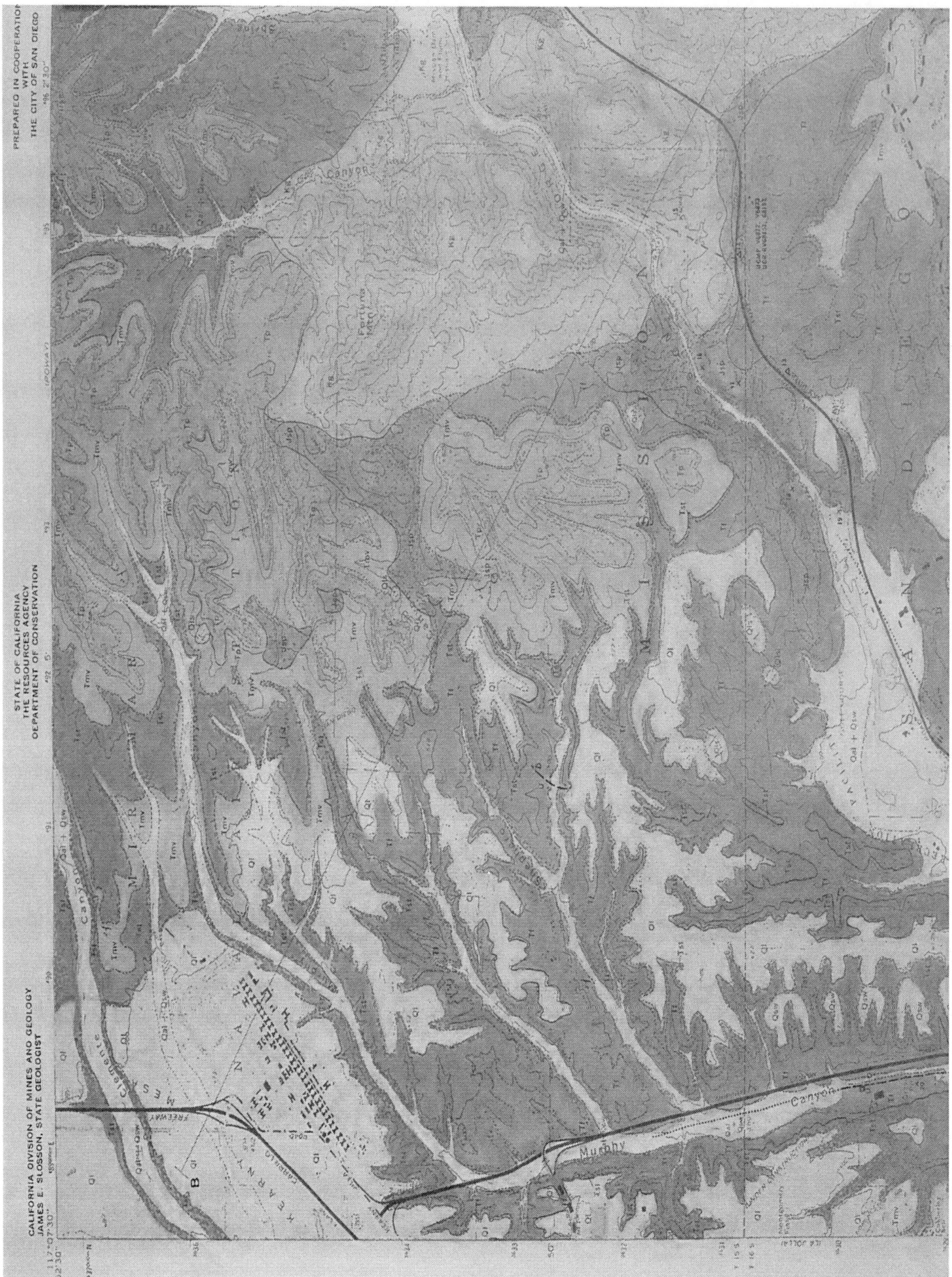
CALIFORNIA DIVISION OF MINES AND GEOLOGY
JAMES E. SLOSSON, STATE GEOLOGIST
STATE OF CALIFORNIA
THE RESOURCES AGENCY
DEPARTMENT OF CONSERVATION
PREPARED IN COOPERATION WITH THE CITY OF SAN DIEGO
FREEWAY
Canyon
Murphy
Canyon

PREPARED IN COOPERATION
WITH
THE CITY OF SAN DIEGO

SAN DIEGO METROPOLITAN AREA
BULLETIN 200, PLATE 3B

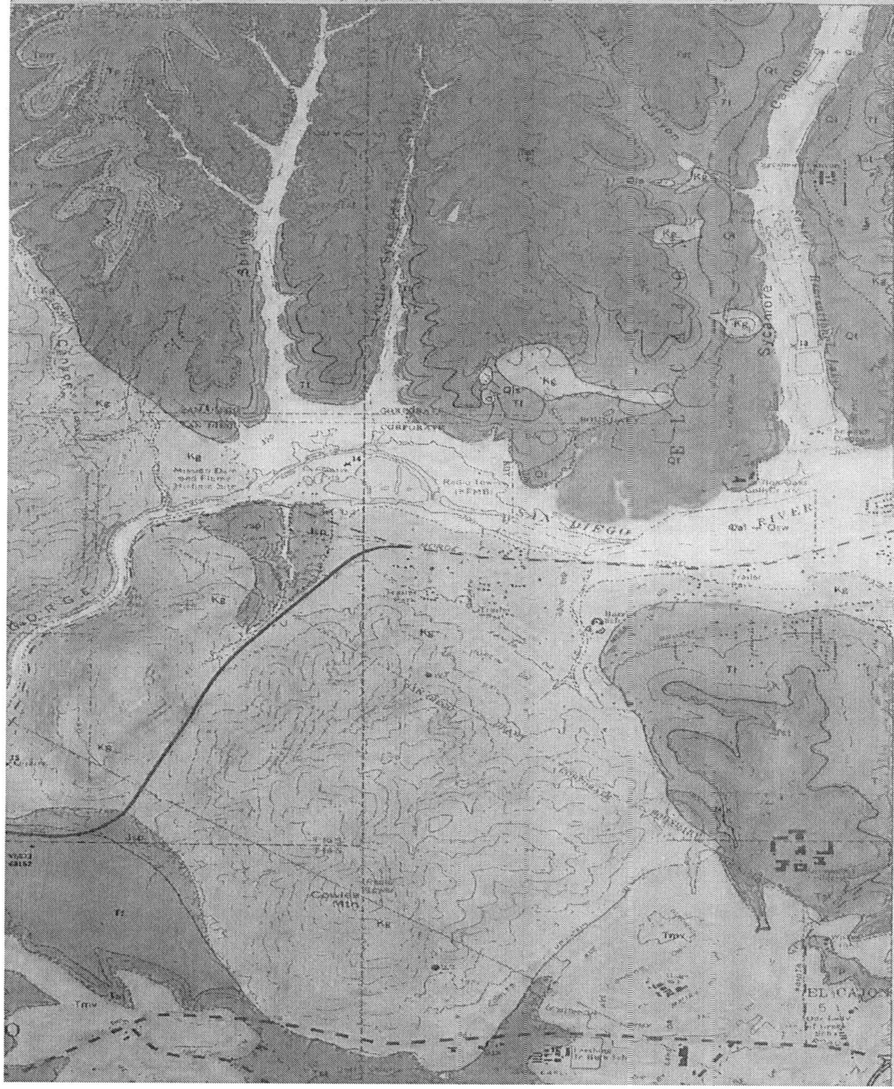

GEOLOGY OF THE LA MESA QUADRANGLE
SAN DIEGO COUNTY, CALIFORNIA

by Michael P. Kennedy and G. L. Peterson

SCALE 1:24 000

CONTOUR INTERVAL 20 FEET
DOTTED LINES REPRESENT 10-FOOT CONTOURS
DATUM IS MEAN SEA LEVEL

1975

EXPLANATION

QUATERNARY

Holocene

Qsw / Qal + Qsw — Alluvium and Slopewash
Qsw, Slopewash. Qal & Qsw, Alluvium and Slopewash undifferentiated.

Qls — Landslide deposits

Pleistocene

Qt — Stream-terrace deposits

Ql — Lindavista Formation

TERTIARY

Pliocene

Tsd — San Diego Formation

Eocene

Tp, Tmv, Tst — Poway Group
Tp, Pomerado Conglomerate. Tmv, Mission Valley Formation. Tst, Stadium Conglomerate. Conglomerate marked by circle pattern.

Tf — Friars Formation
Conglomerate marked by circle pattern.

CRETACEOUS

Kg — Granite rocks
Undifferentiated granitic rocks of the southern California batholith.

JURASSIC

Jsp — Santiago Peak Volcanics

SYMBOLS

Contact
(dashed where approximately located, dotted where concealed)

Fault, showing dip
(dashed where approximately located; dotted where concealed; U, upthrown side; D, downthrown side)

Strike and dip of bedding in sedimentary rocks.

Strike and dip of bedding in metasedimentary rocks.

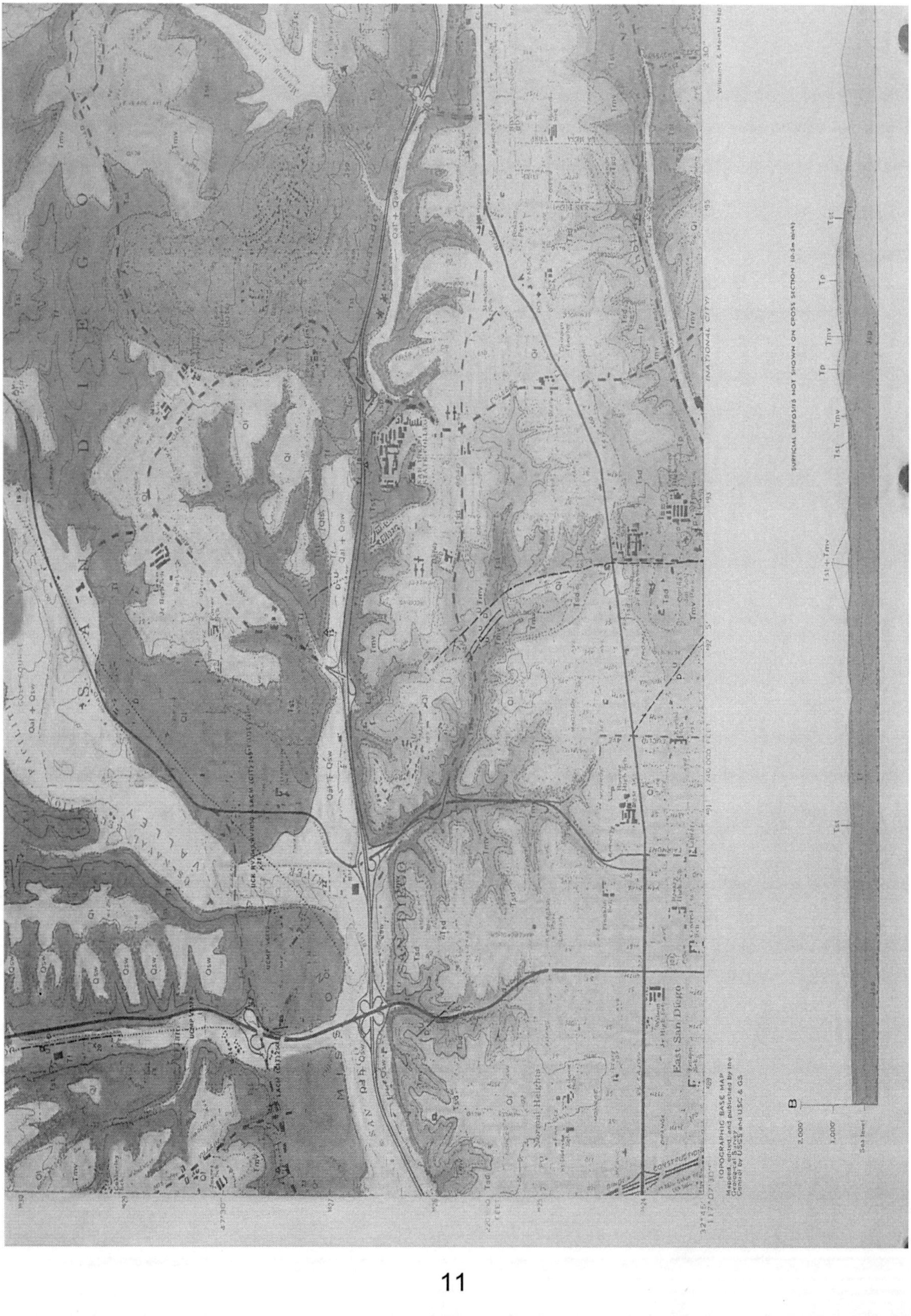
East San Diego
Normal Heights
SAN DIEGO
TOPOGRAPHIC BASE MAP

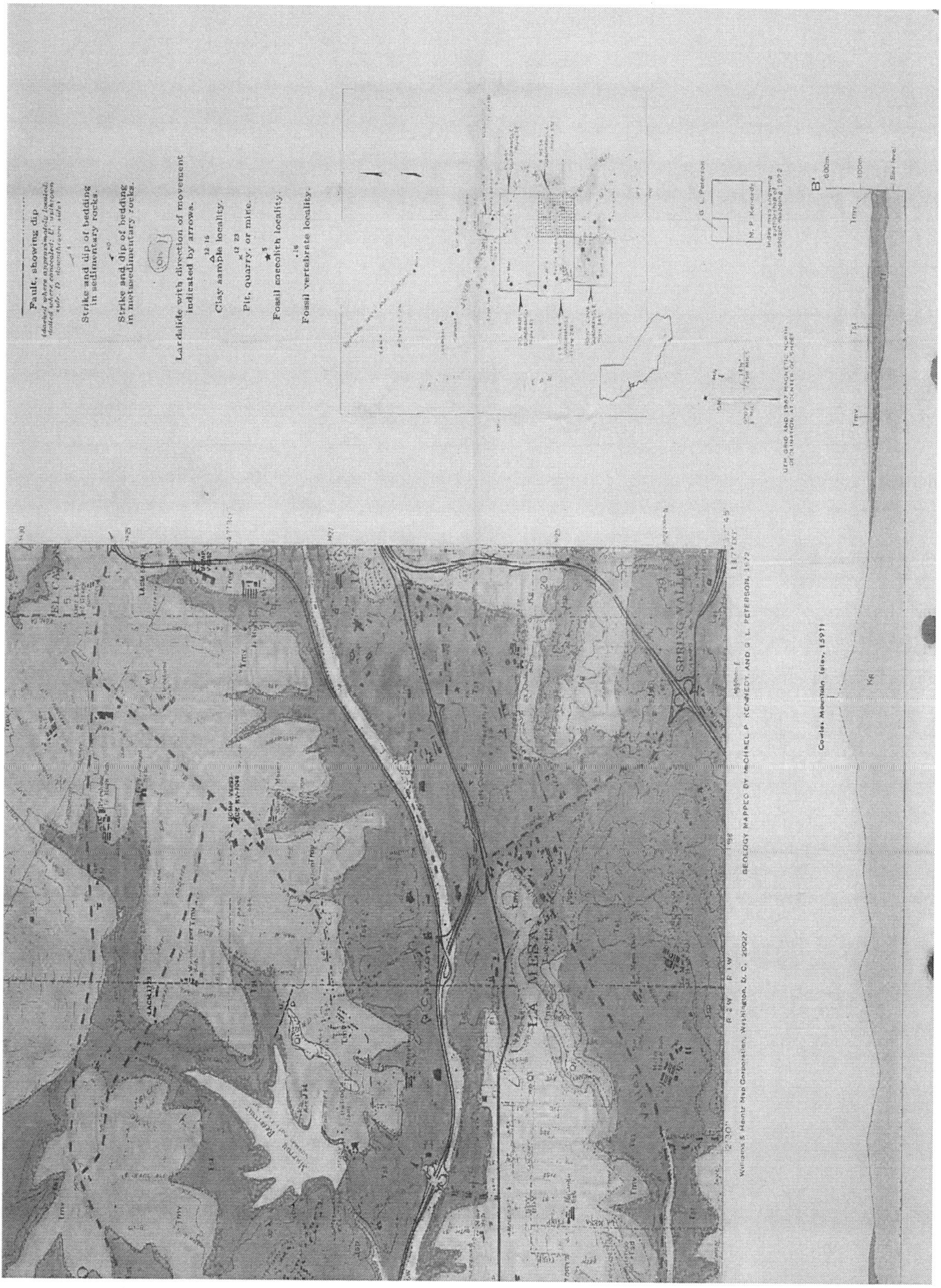
Fault, showing dip
Strike and dip of bedding in sedimentary rocks.
Strike and dip of bedding in metasedimentary rocks.
Landslide with direction of movement indicated by arrows.
Clay sample locality.
Pit, quarry, or mine.
Fossil coccolith locality.
Fossil vertebrate locality.
GEOLOGY MAPPED BY MICHAEL P. KENNEDY AND G. L. PETERSON, 1972
Williams & Heintz Map Corporation, Washington, D. C. 20027
LA MESA
SPRING VALLEY
Cowles Mountain (elev. 1591)
B'

CALIFORNIA DIVISION OF MINES AND GEOLOGY
JAMES E. SLOSSON, STATE GEOLOGIST
STATE OF CALIFORNIA
THE RESOURCES AGENCY
DEPARTMENT OF CONSERVATION
PREPARED IN COOPERATION
WITH
THE CITY OF SAN DIEGO
Canyon

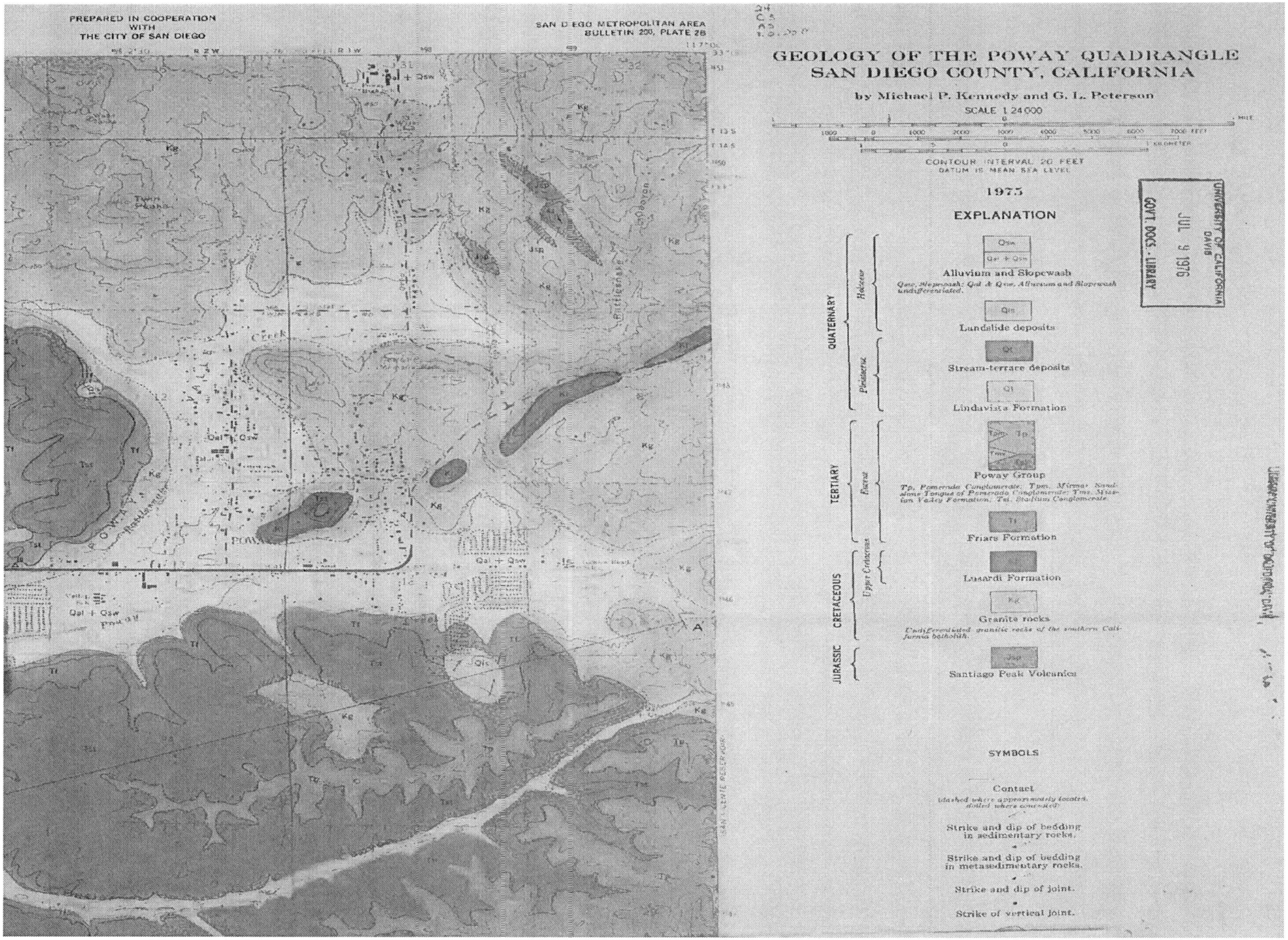
PREPARED IN COOPERATION
WITH
THE CITY OF SAN DIEGO
SAN DIEGO METROPOLITAN AREA
BULLETIN 200, PLATE 28
GEOLOGY OF THE POWAY QUADRANGLE
SAN DIEGO COUNTY, CALIFORNIA
by Michael P. Kennedy and G. L. Peterson
SCALE 1:24 000
CONTOUR INTERVAL 20 FEET
DATUM IS MEAN SEA LEVEL
1975
UNIVERSITY OF CALIFORNIA
DAVIS
JUL 9 1976
GOV'T. DOCS. - LIBRARY
EXPLANATION
QUATERNARY
Holocene
Alluvium and Slopewash
Landslide deposits
Pleistocene
Stream-terrace deposits
Lindavista Formation
TERTIARY
Eocene
Poway Group
Friars Formation
CRETACEOUS
Upper Cretaceous
Lusardi Formation
Granite rocks
JURASSIC
Santiago Peak Volcanics
SYMBOLS
Contact
Strike and dip of bedding in sedimentary rocks.
Strike and dip of bedding in metasedimentary rocks.
Strike and dip of joint.
Strike of vertical joint.
Twin Peaks
Poway
Qal + Qsw

SURFICIAL DEPOSITS NOT SHOWN ON CROSS SECTION
Pomerado Road
Cypress Canyon
Undifferentiated Mesozoic rocks
TOPOGRAPHIC BASE MAP
Sea level

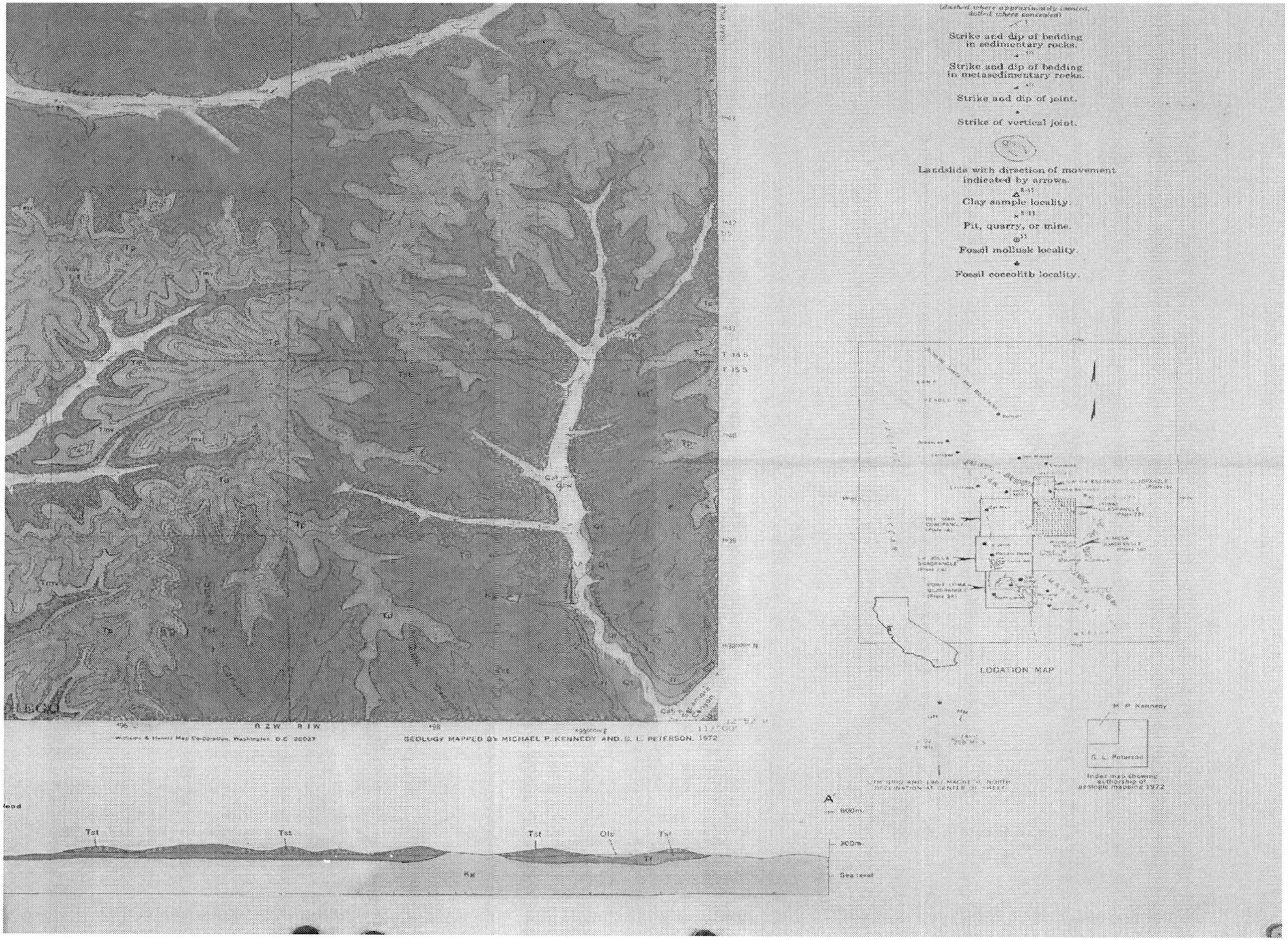
(dashed where approximately located, dotted where concealed)
Strike and dip of bedding in sedimentary rocks.
Strike and dip of bedding in metasedimentary rocks.
Strike and dip of joint.
Strike of vertical joint.
Landslide with direction of movement indicated by arrows.
Clay sample locality.
Pit, quarry, or mine.
Fossil mollusk locality.
Fossil coccolith locality.
LOCATION MAP
M. P. Kennedy
S. L. Peterson
Index map showing authorship of geologic mapping 1972
GEOLOGY MAPPED BY MICHAEL P. KENNEDY AND S. L. PETERSON, 1972
R 2 W
R 1 W
T 14 S
T 15 S
A′
600m.
300m.
Sea level
Tst
Ols
Tf
Kg

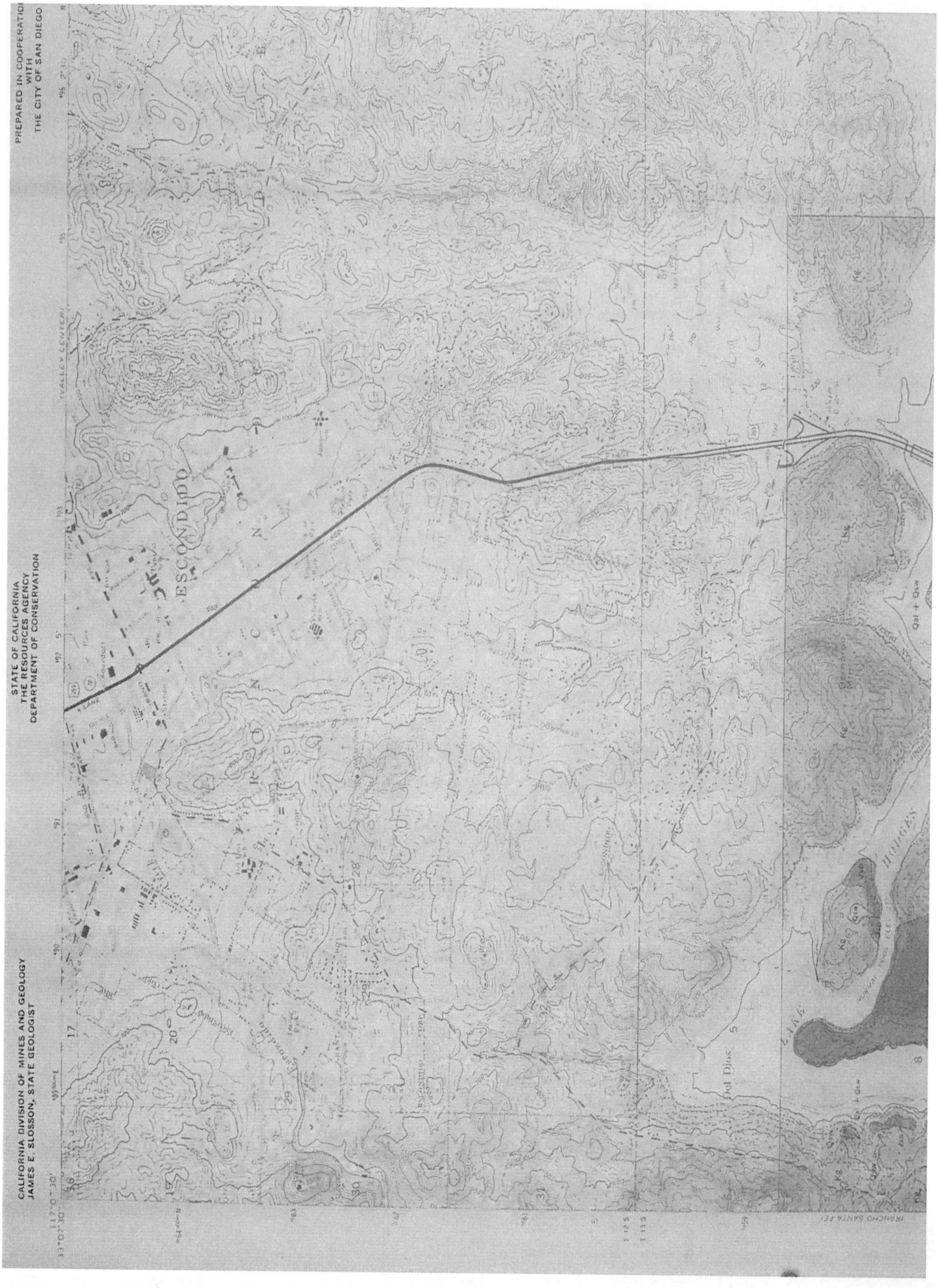
CALIFORNIA DIVISION OF MINES AND GEOLOGY
JAMES E. SLOSSON, STATE GEOLOGIST
STATE OF CALIFORNIA
THE RESOURCES AGENCY
DEPARTMENT OF CONSERVATION
PREPARED IN COOPERATION
WITH
THE CITY OF SAN DIEGO
ESCONDIDO
LAKE
HODGES
Del Dios
Qal + Qsw
RANCHO SANTA FE

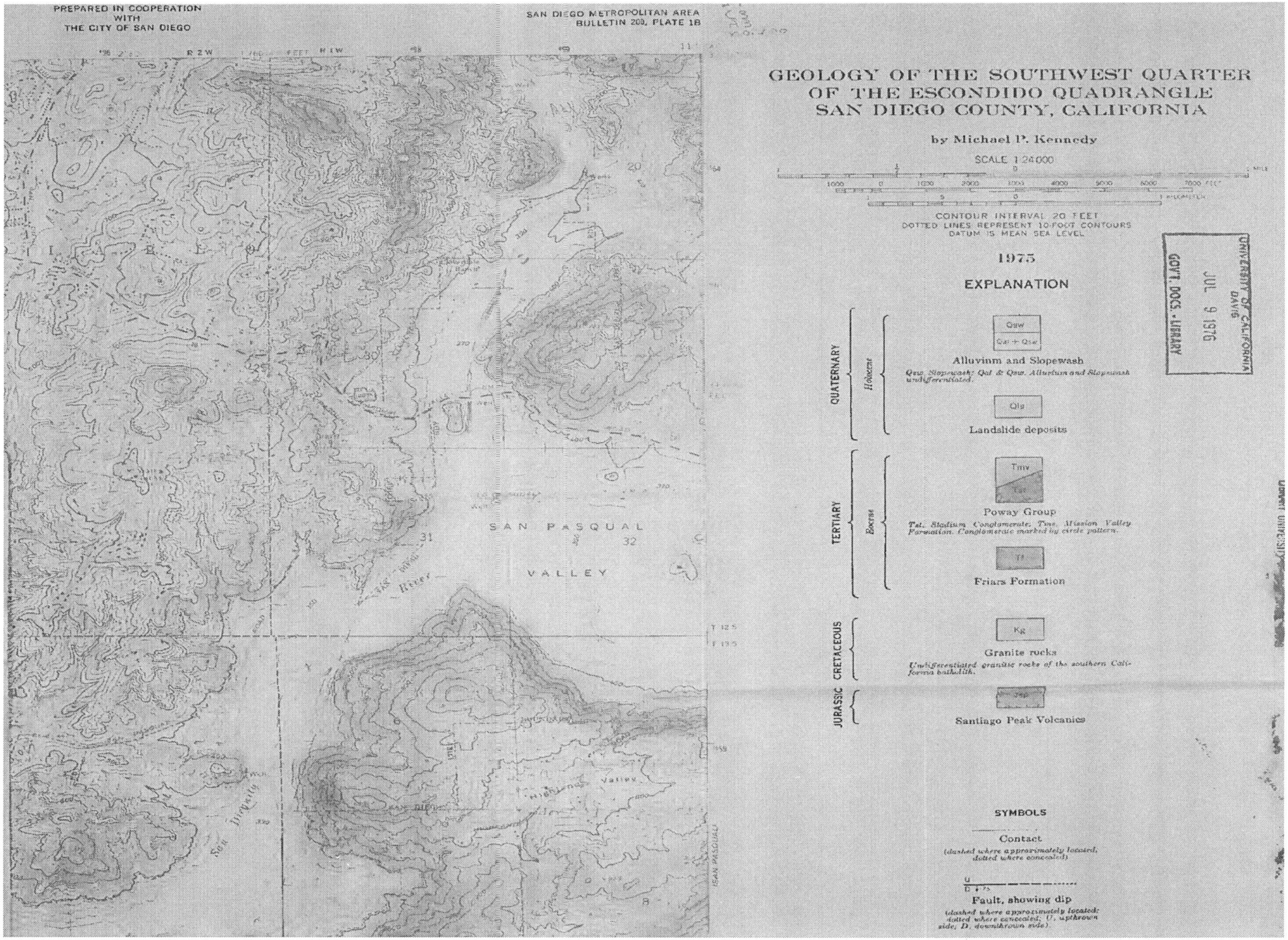
PREPARED IN COOPERATION
WITH
THE CITY OF SAN DIEGO
SAN DIEGO METROPOLITAN AREA
BULLETIN 200, PLATE 1B
GEOLOGY OF THE SOUTHWEST QUARTER
OF THE ESCONDIDO QUADRANGLE
SAN DIEGO COUNTY, CALIFORNIA
by Michael P. Kennedy
SCALE 1:24 000
CONTOUR INTERVAL 20 FEET
DOTTED LINES REPRESENT 10-FOOT CONTOURS
DATUM IS MEAN SEA LEVEL
1975
EXPLANATION
QUATERNARY
Holocene
Qsw
Qal + Qsw
Alluvium and Slopewash
Qsw, Slopewash; Qal & Qsw, Alluvium and Slopewash undifferentiated.
Qls
Landslide deposits
TERTIARY
Eocene
Tmv
Tst
Poway Group
Tst, Stadium Conglomerate; Tmv, Mission Valley Formation. Conglomerate marked by circle pattern.
Tf
Friars Formation
CRETACEOUS
Kg
Granite rocks
Undifferentiated granitic rocks of the southern California batholith.
JURASSIC
Santiago Peak Volcanics
SYMBOLS
Contact
(dashed where approximately located, dotted where concealed)
Fault, showing dip
(dashed where approximately located; dotted where concealed; U, upthrown side; D, downthrown side).
SAN PASQUAL
VALLEY
UNIVERSITY OF CALIFORNIA
DAVIS
JUL 9 1976
GOV'T. DOCS. - LIBRARY

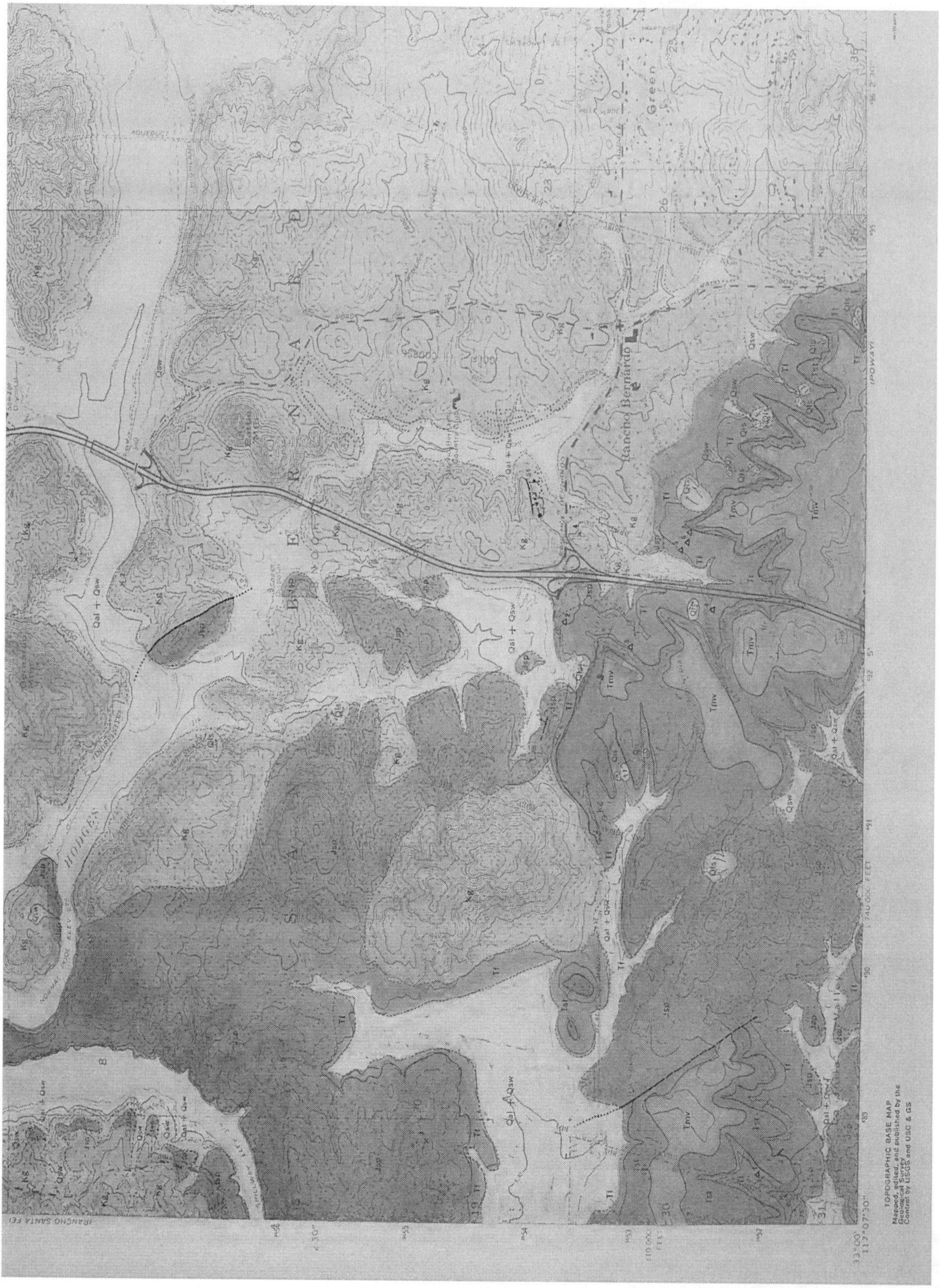
Rancho Bernardo
HODGES
(POWAY)
(RANCHO SANTA FE)
TOPOGRAPHIC BASE MAP
Mapped, edited, and published by the Geological Survey
Control by USGS and USC & GS

Williams & Heintz Map Corporation, Washington, D.C. 20027

GEOLOGY MAPPED BY MICHAEL P. KENNEDY, 1971

Contact
(dashed where approximately located; dotted where concealed)

Fault, showing dip
(dashed where approximately located; dotted where concealed; U, upthrown side; D, downthrown side).

Strike and dip of bedding in sedimentary rocks.

Strike and dip of bedding in metasedimentary rocks.

Strike and dip of joint.

Strike of vertical joint.

Landslide with direction of movement indicated by arrows.

Clay sample locality.

Pit, quarry, or mine.

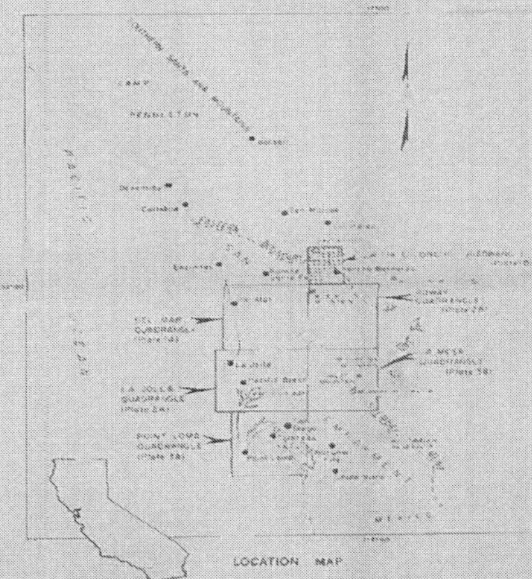

LOCATION MAP

UTM GRID AND 1968 MAGNETIC NORTH DECLINATION AT CENTER OF SHEET